MW00836769

DIFFERENTIAL GEOMETRY

A Geometric Introduction

DIFFERENTIAL GEOMETRY

A Geometric Introduction

David W. Henderson
Cornell University

with writing input from

Daina Taimiņa
University of Latvia

PRENTICE HALL
Upper Saddle River, New Jersey 07458

Library of Congress Cataloging-in-Publication Data

Henderson, David W. (David Wilson), 1939
Differential Geometry : a geometric introduction / David W. Henderson.
p. cm.
Includes bibliographical references (p. -) and index.
ISBN 0-13-569963-0
1. Geometry, Differential. I. Title.
QA641.H474 1998
516.3'6—dc21 97-17042
CIP

Tim Bozik, editorial director
Jerome Grant, editor-in-chief
George Lobell, acquisition editor
Gale Epps, editorial assistant
Kathleen Schiaparelli, executive managing editor
Linda Mihatov Behrens, managing editor
Nancy Gross, editorial assistant
Nicholas Romanelli, production editor
Jayne Conte, cover design
Trudi Pisciotti, manufacturing manager
Alan Fischer, manufacturing buyer
Melody Marcus, marketing manager

© 1998 by Prentice-Hall, Inc.
Simon & Schuster/A Viacom Company
Upper Saddle River, New Jersey 07458

All rights reserved. No part of this book may be reproduced, in any form or by any means, without permission in writing from the publisher.

Portions of this material are based on work supported by the National Science Foundation under Grant No. USE-9155873. Any opinions, findings, and conclusions or recommendations expressed in this material are those of the author and do not necessarily reflect the views of the National Science Foundation.

Printed in the United States of America
10 9 8 7 6 5 4 3 2 1

ISBN 0-13-569963-0

Prentice-Hall International (UK) Limited, London
Prentice-Hall of Australia Pty. Limited, Sydney
Prentice-Hall Canada Inc., Toronto
Prentice-Hall Hispanoamericana, S.A., Mexico
Prentice-Hall of India Private Limited, New Delhi
Prentice-Hall of Japan, Inc., Tokyo
Simon & Schuster Asia Pte. Ltd., Singapore
Editora Prentice-Hall do Brazil, Ltda., Rio de Janeiro

To Daina, Ætna, and the angels

Geometry

Logic can only go so far —
after that I must see-perceive-imagine.
This geometry can help.

I may reason logically thru theorem
and propositions galore,
but only what I perceive is real.

If after studying I am not changed —
if after studying I still see the same —
then all has gone for naught.

Geometry is to open up my mind
so I may see what has always been behind
the illusions that time
and space construct.

Space isn't made of point and line
the points and lines are in the mind.
The physicists see space as curved
with particles that are quite blurred.
And, when I draw, everything is fat
there are no points and that is that.
The artists and the dreamer knows
that space is where an image grows.
For me it's a sea in which I swim
a formless sea of hope and whim.

Thru my fear of Infinity and One
I structure space to confine
my imagination away from the idea
that all is One.

But, I can from this trap escape —
I can see the geometry in which I wander
as but a structure I made to ponder.

I can dare to let go the structures and my fears
and look beyond
to see what is always there to see.

But, to let go, I must first grab on.
Geometry is both the grabbing on
and the letting go.
It is a logical structure
and a perceived meaning —
Q.E.D.'s and "Oh! I see!"'s.
It is formal abstractions
and beautiful contraptions.
It is talking precisely about that
which we know only fuzzily.
But, in the end, and, most of all,
it is seeing-perceiving
the meaning that
I AM.

— David Henderson, 1978

Contents

Preface

> In mathematics, as in any scientific research, we find two tendencies present. On the one hand, the tendency toward *abstraction* seeks to crystallize the *logical* relations inherent in the maze of material that is being studied, and to correlate the material in a systematic and orderly manner. On the other hand, the tendency toward *intuitive understanding* fosters a more immediate grasp of the objects one studies, a live *rapport* with them, so to speak, which stresses the concrete meaning of their relations.
>
> As to geometry, in particular, the abstract tendency has here led to the magnificent systematic theories of Algebraic Geometry, of Riemannian Geometry, and of Topology; these theories make extensive use of abstract reasoning and symbolic calculation in the sense of algebra. Notwithstanding this, it is still as true today as it ever was that intuitive understanding plays a major role in geometry. And such concrete intuition is of great value not only for the research worker, but also for anyone who wishes to study and appreciate the results of research in geometry.
>
> — David Hilbert [**SE**: Hilbert]

These words, written in 1934 by David Hilbert, the "father of Formalism," are from the Preface to *Geometry and the Imagination.*

The formalisms of differential geometry are considered by many to be among the most complicated and inaccessible of all the formal systems in mathematics. It is probably fair to say that most mathematicians do not feel comfortable with their understanding of differential geometry. In addition, there is little agreement about which formalisms to use or how to describe them, with the result that the starting definitions, notations and analytic descriptions vary widely from text to text. What all of these different approaches have in common are underlying geometric intuitions of the basic notions such as straightness (geodesic), smooth, tangent, curvature, and parallel transport.

In this book we will study a foundation for differential geometry based not on analytic formalisms but rather on these underlying geometric intuitions. This foundation should be accessible to anyone with a flexible geometric imagination. It may then be possible that this foundation will serve as a common starting point for the various analytic formalizations. We will explore some of these analytical formalisms. In addition, this geometric foundation relates more directly with our actual experiences of curves and surfaces both in the physical world and in the context of computer graphics.

I invite the reader to explore the basic ideas of differential geometry. I am interested in conveying a different approach to mathematics, stimulating the reader to a broader and deeper experience of mathematics. Active participation with these ideas, including exploring and writing about them, will give the reader a broader context and experience, which is vital for anyone who wishes to understand differential geometry at a deeper level. More and more of the formal analytical aspects of differential geometry have now been mechanized, and this mechanization is widely available on personal computers, but the experience of meaning in differential geometry is still a human enterprise that is necessary for creative work.

I believe that mathematics is a natural and deep part of human experience and that experiences of meanings in mathematics are accessible to everyone. Much of mathematics is not accessible through formal approaches except to those with specialized learning. However, through the use of non-formal experience and geometric imagery, many levels of meaning in mathematics can be opened up in a way that more human beings can experience and find intellectually challenging and stimulating.

This text builds on a foundation of intuitive geometric ideas and then ties them into the formalisms of extrinsic and intrinsic differential geometry. The first chapter is an extensive collection of examples of surfaces which are discussed as much as can be done using elementary techniques and geometric intuition. Many of the concepts in the text are introduced in Chapter 1. Throughout the text there is an emphasis on looking at curves and surfaces in as many different ways as possible but with a particular emphasis on intrinsic, coordinate-free approaches in order to highlight the geometry.

The book is written for undergraduate mathematics majors and thus assumes of the reader a corresponding level of interest and mathematical sophistication. Previous experience with multivariable calculus and linear algebra are strongly recommended. There is more material in this text than I cover in one semester, so the instructor can choose to leave out certain topics. To assist in this process some problems or parts of problems are preceded by an asterisk (*), indicating that they are not essential for the remainder of the text.

There are some results in this text which (insofar as I know) have never before appeared in print. These include the annular hyperbolic plane (that I learned from William Thurston, see Problems **1.8** and **5.7**); the use of zooming and fields of view in defining smoothness (see the beginnings of Chapters 2 and 3, especially Problems **2.1**, **2.2**, and **3.1**); and the Ribbon Test for a geodesic (Problems **3.4** and **7.5**).

There is a unique problem-centered approach in the presentation of this material. The main geometric notions, both theory and concepts, are introduced through problems which are designed to give students an opportunity to experience their own meanings in the material. This is similar to the approach used successfully in the author's *Experiencing Geometry on Plane and Sphere* (Prentice Hall, 1996). Two discussions describing different aspects of this approach can be found in [David Henderson, "I Learn Mathematics From My Students — Multiculturalism in Action," *For the Learning of Mathematics*, v.**16**, n.2, pp.46-52] and [Jane-Jane Lo, Kelly Gaddis and David Henderson, "Learning Mathematics Through Personal Experiences: A Geometry Course in Action," *For the Learning of Mathematics*, v.**16**, n.1, pp.34-40].

Useful Supplements

Instructors may obtain from the publisher an Instructor's Solution Manual (containing possible solutions for each problem) by sending a request via e-mail George_Lobell@prenhall.com or call 1-201-236-7407.

Those readers who have access to computer systems running Maple©, Mathematica©, Derive©, or similar software can use these systems to facilitate gaining geometric intuition of the concepts of differential geometry. In Appendix C I have included several computer exercises for Maple, and these and additional scripts are also available for downloading on-line at ftp://math.cornell.edu/pub/Henderson/diff_geom. I will also include at this site additonal information and updates as they become available.

Acknowledgments

I acknowledge my debt to all the students who have attended my differential geometry courses. Without them this book would have been an impossibility.

I started writing problems such as those that appear in this book while teaching differential geometry in the spring of 1992. Again in the spring of 1994, I wrote more problems and used them together with a published textbook for the course. In the spring of 1995 I taught the course using only my problems and altered them and extended them as we went along. Finally, the first preliminary version of this text was available in photocopy form in the fall of 1995. It was used by Brian Mortimer in his differential geometry class at Carleton University in Ottawa, Canada, and by James West in his differential geometry course at Cornell University. Both of these mathematicians gave me valuable feedback. In addition, I received many helpful suggestions and comments from the reviewers of the Fall 1995 version and later the Fall 1996 version. These reviewers were Brian Mortimer, Bruce Piper (Rensselear Polytechnic Institute), Nicola Garofalo (Purdue University), George C. Johnson (University of California at Berkeley), Steven L. Kent (Youngstown State University), Larry Cusick (California State University at Fresno), Bruce Hughes (Vanderbilt University), and several anonymous reviewers.

In addition, I received valuable feedback from Jane-Jane Lo, Dexter Luthulli, Cathy Stenson, and Alex Tsow. In the spring of 1997 I used a near final version of the text in a course at Cornell. From these students, instructors, and others, I received encouragement and much valuable feedback that resulted in what I consider to be a better book. Cathy Stenson also wrote the Maple© computer scripts.

In October 1995, I gave a copy of the preliminary text to Daina Taimiņa, faculty member of the Faculty of Mathematics and Physics at the University of Latvia in Riga. Much of the final rewriting and extending of this text was completed with her assistance during my two-month visit to Latvia in the summer of 1996 and her visit to Ithaca during the spring of 1997.

The entire production of the manuscript (typing, formatting, drawings, and final layout and typesetting) has been accomplished using Ami-Pro, an integrated word-processing software.

Finally, I wish to thank George Lobell, Senior Editor at Prentice Hall, for his contagious enthusiasm and for the vision with which he shepherded this book through the publication process.

Ithaca, NY, June 1997

David W. Henderson

How to Use This Book

Do not just pay attention to the words;
Instead pay attention to meanings behind the words.
But, do not just pay attention to meanings behind the words;
Instead pay attention to your deep experience of those meanings.
— Tenzin Gyatso, The Fourteenth Dalai Lama[†]

This quote expresses the philosophy upon which this book is based. This book will present you with a series of problems. You should explore each question and write out your thinking in a way that can be shared with others. By doing this you will be able to actively develop ideas prior to passively reading or listening to comments of others. When working on the problems, you should be open-minded and flexible and let your thinking wander. Some problems will have short, fairly definitive answers, and others will lead into deep areas of meaning which can be probed almost indefinitely. You should not accept anything just because you remember it from school or because some authority says it is good. Insist on understanding (or seeing) why it is true or what it means to you. Pay attention to **your** deep experience of these meanings.

You should think about the problems and express your thinking about them even when you know you cannot complete the problem. This is important because:

- It helps build self confidence.
- You will see what your real difficulties are.
- When you see a solution or proof later, then you will more likely see it as answering a question that you have.

It is important for you to keep in mind that there is more than one correct solution. There are many different ways of solving the problems — as many as there are ways of understanding the problems. *Insist on understanding* (or seeing) why it is true or what it means to you. Everyone understands things in a different way, and one person's "obvious" solution may not work for you. However, it is helpful to talk with others

[†]From an unpublished lecture in London, April, 1984. Used here with permission.

—listen to their ideas and confusions and then share your ideas and confusions with them.

Also, some of the problems are difficult to visualize in your head. Make models, draw pictures, use rubber bands on a ball, use scissors and paper — play!

Throughout the text there is an emphasis on looking at curves and surfaces in as many different ways as possible but with a particular emphasis on intrinsic, coordinate-free approaches in order to highlight the geometry. Those readers who wish to avoid local coordinates may, in general, do so by leaving out the problems and sections that refer to them. From the beginning through Problem **4.7** local coordinates are used only as examples. Local coordinates and the associated formalisms are needed in a crucial way only in Problem **4.8** and in the problems following Problem **6.1**.

Those readers who have access to computer systems running Maple©, Mathematica©, Derive©, or similar software can use these systems to facilitate gaining geometric intuition and imagination of the concepts of differential geometry. In Appendix C we have included several computer exercises for Maple© and these and additional scripts are also available for downloading on-line at

ftp://math.cornell.edu/Henderson/diff_geom.

Similar scripts for other programs should be easily constructed. These exercises are labeled according to the problem in the text to which they are most applicable. They are also referenced at appropriate points in the text. However, the current state-of-the-art for generally available computer graphing programs is not capable of producing what would be the most useful displays. For example, it is not currently possible, with widely useable programs, to view a curve on a surface and to use the mouse to dynamically move a point along the curve and see displayed the three curvature vectors—intrinsic (geodesic), extrinsic, and normal. I hope that interested readers will add to the available collection of scripts by sending to me (dwh@math.cornell.edu) their scripts or URL's to where their scripts are available on the WWW. Please also use this same e-mail address to send any comments about the book or its use.

Chapter 1
Surfaces and Straightness†

In the first six chapters of this book our study of differential geometry focuses on curves and surfaces. In later chapters we will see how to extend the results about surfaces to higher dimensional manifolds (the higher dimensional analogues of surfaces), especially to our physical universe which is a 3-dimensional (or 4-dimensional, if you include time) manifold.

In this chapter we will begin our study by examining a diverse collection of surfaces which will serve as examples throughout the remainder of the book. We will investigate each surface as much as we can without bringing in the differential notions of calculus. For each surface, starting with the plane, we will say what we can about what it means to be straight on the surface.

We begin with a question that encourages you to explore deeply a concept that is fundamental to all that will follow: We ask you to build a notion of straightness for yourself rather than accept a certain number of assumptions about straightness. Although difficult to formalize, straightness is a natural human concept.

PROBLEM 1.1. When Do You Call a Line Straight?

Look to your experiences. It might help to think about how you would explain straightness to a 5-year-old (or how the 5-year-old might explain it to you!). If you use a "ruler," how do you know if the ruler is straight? How can you check it? What properties do straight lines have that distinguish them from non-straight lines?

Think about the question in four related ways:

†A small portion of this chapter is taken (somewhat revised) from the author's *Experiencing Geometry on Plane and Sphere* [**Tx**: Henderson]. It is used here with the permission of the publisher, Prentice-Hall, Inc.

a. *How can you check in a practical way if something is straight—without assuming that you have a ruler, for then we will ask, "How can you check that the ruler is straight?"*

b. *How do you construct something straight—lay out fence posts in a straight line, or draw a straight line?*

c. *What symmetries does a straight line have? A symmetry of a geometric figure is a transformation (such as reflection, rotation, translation, or composition of them) which preserves the figure. For example, the letter* "T" *has reflection symmetry about a vertical line through its middle, and the letter* "Z" *has rotation symmetry if you rotate it half a revolution about its center.*

d. *Can you write a definition of "straight line"?*

Suggestions

Look at your experience. At first, you will look for examples of physical world (or natural) straightness. Go out and actually try walking along a straight line and then along a curved path; try drawing a straight line and checking that a line already drawn is straight.

Look for things that you call "straight." Where do you see straight lines? Why do you say they are straight? Look for both physical lines and non-physical uses of the word "straight." You are likely to bring up many ideas of straightness. It is necessary then to think about what is common among all of these "straight" phenomena.

As you look for properties of straight lines that distinguish them from non-straight lines, you will probably remember the following statement (which is often taken as a definition in high school geometry): *A line is the shortest distance between two points.* But can you ever measure the lengths of all the paths between two points? How do you find the shortest path? If the shortest path between two points is in fact a straight line, then is the converse true? Is a straight line between two points always the shortest path? We will return to these questions later in this chapter.

A powerful approach to this problem is to think about lines in terms of symmetry. Two symmetries of lines in the plane are:

- Reflection symmetry in the line, also called bilateral symmetry—reflecting (or mirroring) an object over the line (Figure 1.1).

Figure 1.1. Bilateral symmetry.

- Half-turn symmetry—rotating 180° about any point on the line (Figure 1.2).

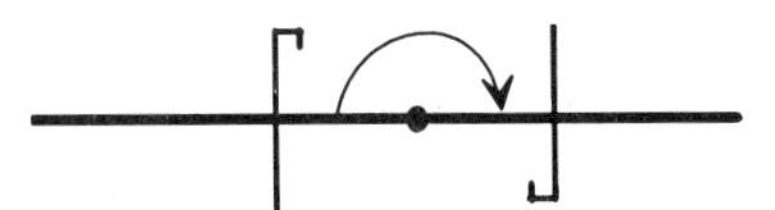

Figure 1.2. Half-turn symmetry.

Although we are focusing on a symmetry of the line in each of these examples, notice that the symmetry is not a property of the line by itself but includes the line and the space around the line. The symmetries preserve the local environment of the line. Notice how in reflection and half-turn symmetry the line and its local environment are both part of the symmetry action. This relationship between them is integral to the action. In fact, reflection in the line does not move the line at all but exhibits a way in which the spaces on the two sides of the line are the same.

Try to think of other symmetries as well (there are quite a few). Some symmetries hold only for straight lines, while some work with other curves too. Try to determine which ones are specific to straight lines and why. Also think of practical applications of these symmetries for constructing a straight line or for determining if a line is straight.

How Do You Construct a Straight Line?

As for how to construct a straight line, one method is simply to fold a piece of paper; the edges of the paper needn't even be straight. This utilizes symmetry (can you see which one?) to produce the straight line. Carpenters also use symmetry to determine straightness—they put two boards face to face and plane the edges until they look and feel straight.

They then turn one board over so the planed edges are touching, then hold the boards up to the light. If the edges are not straight, there will be gaps between the boards through which light will shine. (See Figure 1.3.)

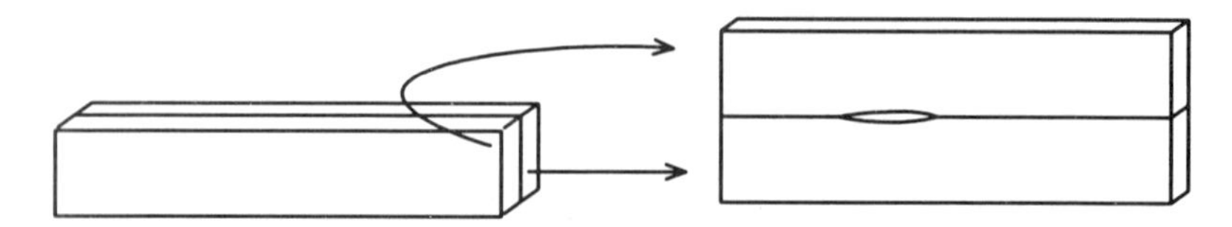

Figure 1.3. Carpenter's method for checking straightness.

To grind an extremely accurate flat mirror, the following technique is sometimes used: Take three approximately flat pieces of glass and put pumice between the first and second pieces and grind them together. Then do the same for the second and the third pieces and then for the third and first pieces. Repeat many times and all three pieces of glass will become very accurately flat. (See Figure 1.4.) Do you see why this process works? What does this have to do with straightness?

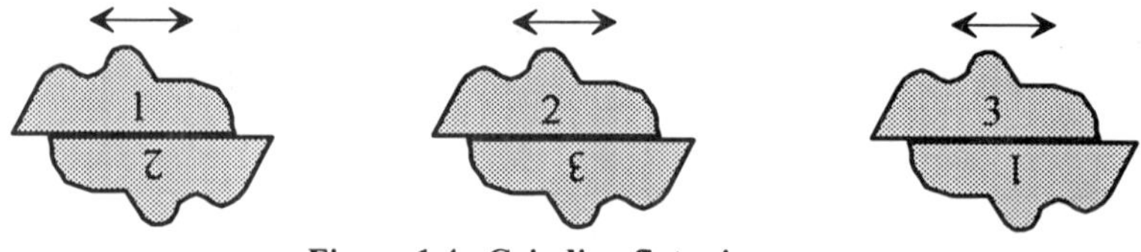

Figure 1.4. Grinding flat mirrors.

Imagine walking (or actually do it!) while pulling a long silk thread with a small stone attached. When will the stone follow along your path? Why? To illustrate this phenomenon, consider how a fallen water skier can be rescued. The boat passes by the skier at a safe distance in a straight path, and the tow rope follows the path of the boat. The boat then turns in an arc in front of the skier. Since the boat is no longer following a straight path, the tow rope will move in toward the fallen skier. In these two examples there is a stretched thread or rope which follow the shortest distance, thus these illustrate the property that straight lines are locally the shortest distance. See, also, the section Is "Shortest" Always "Straight"? later in this chapter.

Another idea to keep in mind is that straightness must be thought of as a local property. Part of a line can be straight even though the whole line may not be. For example, if we agree that this line is straight,

and then we add a squiggly part on the end, like this:

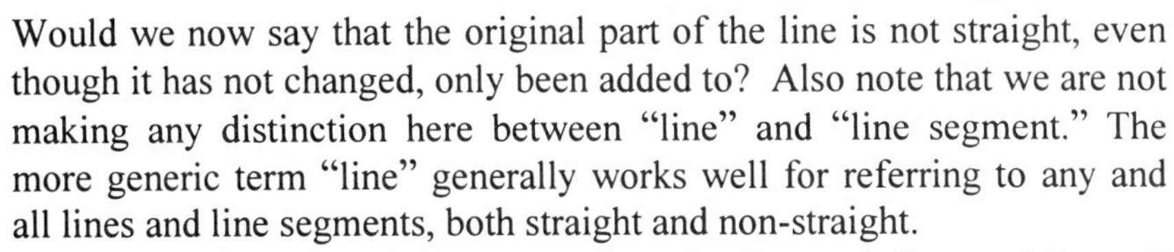

Would we now say that the original part of the line is not straight, even though it has not changed, only been added to? Also note that we are not making any distinction here between "line" and "line segment." The more generic term "line" generally works well for referring to any and all lines and line segments, both straight and non-straight.

Can we use any of the symmetries of a line to define straightness? What symmetries does a straight line have? How do they fit with the examples that you have found and those mentioned above?

Returning to one of the original questions, how would we construct a straight line? One way would be to use a "straight edge" — something that we accept as straight. Notice that this is different from the way that we would draw a circle. When using a compass to draw a circle, we are not starting with a figure that we accept as circular; instead, we are using a fundamental property of circles that the points on a circle are a fixed distance from the center. Can we use the symmetry properties of a straight line to construct a straight line? Is there a tool (serving the role of a compass) which will draw a straight line? For an interesting discussion of this question see *How to Draw a Straight Line: A Lecture on Linkages* (1877) [**Z**: Kempe, p. 12] which shows the apparatus for drawing a straight line that is pictured in Figure 1.5. See [**SE**: Hilbert, pp. 272-3] for another discussion of this topic. The discovery of this linkage about 1870 is variously attributed to the French army officer, M. Peaucellier, and to Lippman Lipkin, who lived in Lithuania. (See Kempe and Hilbert and Phillip Davis' delightful little book *The Thread* [**Z**: Davis, Chapter IV].)

Think about and formulate some answers to these questions before you read any further. You are the one laying down the definitions. Do not take anything for granted unless you see why it is true. No answers are predetermined. You may come up with something that we have never imagined. Consequently, it is important that you persist in following your own ideas.

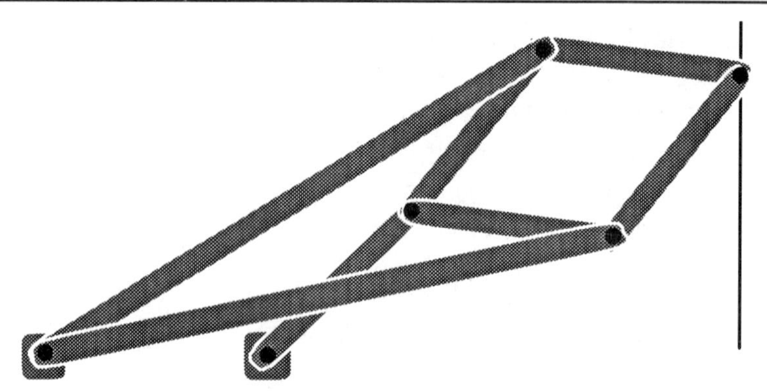

Figure 1.5. Apparatus for drawing a straight line.

Local (and Infinitesimal) Straightness

Previously, you saw how a straight line has reflection-in-the-line symmetry and half-turn symmetry: One side of the line is the same as the other. But, as pointed out above, straightness is a local property in that whether a segment of a line is straight depends only on what is near the segment and does not depend on anything happening away from the line. Thus each of the symmetries must be able to be thought of (and experienced) as applying only locally. This will become particularly important later when we investigate straightness on the cone and cylinder. For now, it can be experienced in the following way:

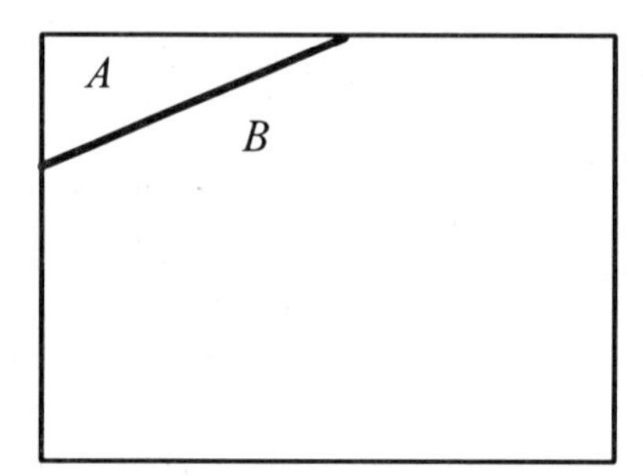

Figure 1.6. Reflection symmetry is local.

When a piece of paper is folded not in the center (like in Figure 1.6), the crease is still straight even though the two sides (A and B) of the crease on the paper are not the same. So what is the role of the sides when we are checking for straightness using reflection symmetry? Think about what is important near the crease in order to have reflection symmetry.

When we talk about straightness as a local property, you may bring out some notions of scale. For example, if one sees only a small portion of a very large circle, it will be indistinguishable from a straight line. This can be experienced easily on many of the modern graphing programs for computers. Also, a microscope with a zoom lens will provide an experience of zooming. For some curves, if one "zooms in" on any point of the curve, eventually the curve will be indistinguishable from a straight line segment. (See Figure 1.7.)

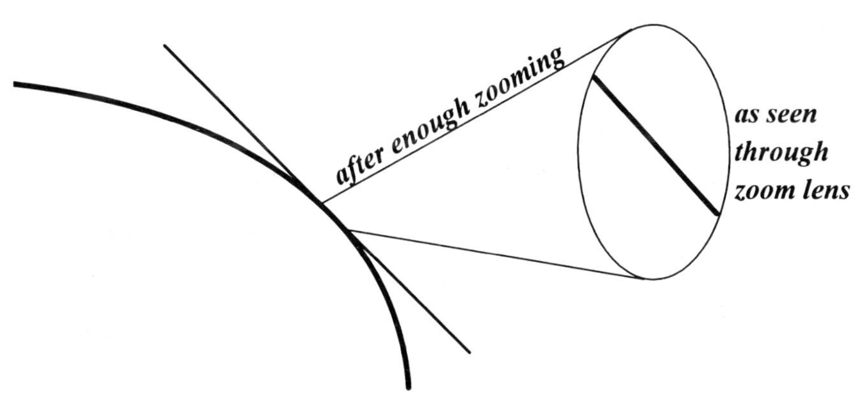

Figure 1.7. Infinitesimally straight.

We call such a curve *infinitesimally straight*. In Chapter 2 we show that this is equivalent to the more standard term, *differentiable*. We also show that a curve is *smooth* (or *continuously differentiable*) if and only if it is *uniformly infinitesimally straight* in the sense made clear in Chapter 2. When the curve is parametrized by arclength this is equivalent to the curve having a continuously defined velocity vector at each point.

In contrast, we can say that a curve is *locally straight at a point* if that point has a neighborhood that is straight. In the physical world the usual use of both *smooth* and *locally straight* are dependent on the scale at which they are viewed. For example, we may look at an arch made out of wood that at a distance appears as a smooth curve (Figure 1.8a); then as we move in closer we see that the curve is made by many short straight pieces of finished (planed) boards (Figure 1.8b), but when we are close enough to touch it, we see that its surface is made up of smooth waves or ripples (Figure 1.8c), and under a microscope we see the non-smoothness of numerous twisting fibers (Figure 1.8d).

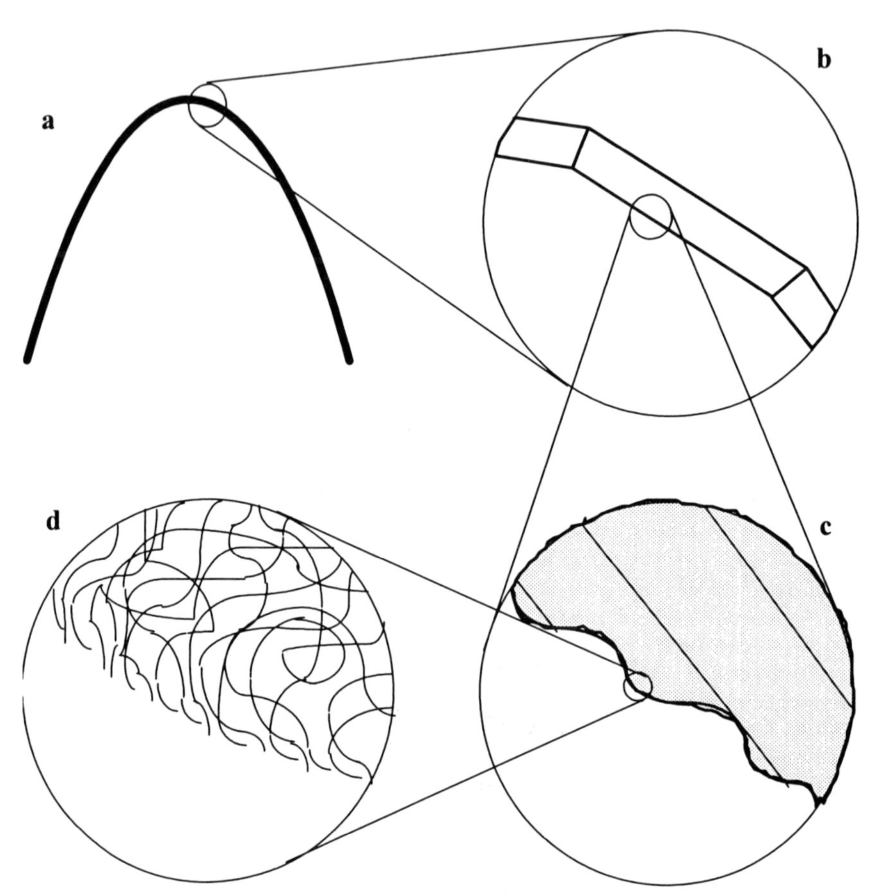

Figure 1.8. Straightness and smoothness depend on the scale.

Problem 1.2. Intrinsic Straight Lines on Cylinders

Take a piece of paper and draw a straight line on the paper. Now bend the paper (without stretching it) so that the line is no longer straight in 3-space. We express this situation by saying that the line is

extrinsically not straight; but is the line straight intrinsically on the sheet of paper? That is, if you consider the paper to be the universe then is the line straight in that universe? Or, consider a 2-dimensional bug which crawls on the surface of the paper such that the bug has no awareness of any space off the surface of the paper and is not influenced by gravity. Will the bug experience the line on the paper as straight?

a. *Argue that distances (as measured along the surface of the bent paper) and angles have not changed and thus that the bent paper will intrinsically have the same geometric properties as a flat piece of paper. Argue that the bug would experience the line on the paper as straight. Argue that the bug would experience the line as having the same (local and intrinsic) symmetries as straight lines on the plane.*

The important thing to remember here is to **think in terms of the surface, not in 3-space**. Always try to imagine how things would look from the bug's point of view. A good example of how this type of thinking works is to look at an insect called a water strider. The water strider walks on the surface of a pond and has a very 2-dimensional perception of the world around it — to the water strider, there is no up or down; its whole world consists of the plane of the water. The water strider is very sensitive to motion and vibration on the water's surface, but it can be approached from above or below without its knowledge. If you find a pond with water striders you can actually, by moving slowly (so as not to disturb the surface of the water with air currents), touch the water strider with your finger. Hungry birds and fish can also take advantage of this 2-dimensional perception. This is the type of thinking needed to adequately visualize the intrinsic geometric properties of any surface.

This leads us to consider the concept of ***intrinsic***, or ***geodesic***, ***curvature*** versus ***extrinsic curvature***. An outside observer in 3-space looking at the bent paper will see the line drawn on the paper as curved—that is, the line exhibits *extrinsic* curvature. But relative to the surface (*intrinsically*), the line has no *intrinsic curvature* and thus is straight. **Be sure to understand this difference**. Lines which are intrinsically straight on a surface are often called ***geodesics***.

All symmetries (such as reflections and half-turns) must be carried out intrinsically, or from the bug's point of view. There will in general not be extrinsic symmetries. For example, on a cylinder there is no extrinsic reflection symmetry except along or perpendicular to one of the generators of the cylinder.

It is natural for you at first to have some difficulty experiencing straightness on surfaces other than the plane, and that consequently you will start to look by looking at the curves on surfaces as 3-dimensional objects. Imagining that you are a 2-dimensional bug walking on the surface emphasizes the importance of experiencing straightness and will help you to shed your limiting extrinsic 3-dimensional vision of the curves on a bent surface. Ask yourself: What does the bug have to do, when walking on a cylindrical surface, in order to walk in a straight line? How can the bug check if it is going straight?

b. *What lines are straight with respect to the surface of a cylinder? Why? Why not? Have you listed all of them? How do you know? If you intersect a cylinder by a flat plane and unroll it, what kind of curve do you get? Is it ever straight?*

Rolling a piece of paper into a cylinder does not change the ***local*** intrinsic geometry, and thus the notions of symmetry should still apply locally and intrinsically for a geodesic on the surface. Thus a helix on a cylinder locally and intrinsically has the two types of reflection symmetry, half-turn symmetry, and rigid-motion-along-itself symmetry. Note that reflection symmetry does not hold globally (that is, as symmetries of the whole cylinder) and does not hold extrinsically (that is, an ordinary extrinsic mirror will not produce symmetry on a helix even locally).

Make paper models, but consider the cylinder as continuing indefinitely with no top or bottom. Again, imagine yourself as a bug whose whole universe is the surface of the cylinder. As the bug crawls around on the surface, what will the bug experience as straight?

Lay a stiff ribbon or straight strip of paper on a cylinder. Convince yourself that it will follow a straight line with respect to the surface. Also, convince yourself that straight lines on the cylinder, when looked at locally and intrinsically, have the same symmetries as on the plane.

Rolling a piece of paper into a cylinder does not change the ***local*** intrinsic geometry but it does change the ***global*** intrinsic geometry. For example, on the cylinder there is a ***closed geodesic*** which returns to its starting point (can you find one?) and this is impossible on the plane. Also note that there is more than one geodesic joining every pair of points on a cylinder.

c. *How many geodesics join two points on a cylinder? How can you find these geodesics? On a cylinder, can a geodesic ever intersect itself? Is an intrinsic straight line on the cylinder always the shortest distance? Is the shortest distance always straight? Why?*

As you begin to explore these questions, it is likely that many other related geometric ideas will arise. Do not let seemingly irrelevant excess geometric baggage worry you. Often, you will find yourself getting lost in a tangential idea, and that is understandable. Ultimately, however, the exploration of related ideas will give you a richer understanding of the scope and depth of the problem. There are several important things to keep in mind while you are working on this problem. First, **you must make models**. You may find it helpful to make models using transparencies. Second, you must think about lines on the cylinder in an intrinsic way—always look at things from a bug's point of view. We are not interested in what is happening in 3-space; only what you would see and experience if you were restricted to the surface of a cylinder. Third, remember that if you cut the cylinder and lay it flat on the plane, then paths that were geodesics on the cylinder will become straight lines on the plane.

Here are some activities that you can try, or visualize, to help you experience what are the geodesics on surfaces. However, it is better for you to come up with your own experiences.

- Stretch something elastic on the surface. It will stay in place along a geodesic, but it will not stay on a curved path if the surface is slippery. Here, the elastic follows a path that is approximately the shortest since a stretched elastic always moves so that it will be shorter. Using the shortest distance criterion directly is not a good way to check for straightness because one cannot possibly measure all paths. But, it serves a good purpose here.

- Roll a cylinder (or other "rollable" surface) on a straight chalk line. The chalk will mark the line of contact on the cylinder and it will be a geodesic.

- Take a stiff ribbon or strip of paper that does not stretch, and lay it "flat" on the surface. It will only lie properly along a geodesic. Do you see how this property is related to local symmetry? This is sometimes called the *Ribbon Test.* (See Problem **3.4**.)

- The feeling of turning and "non-turning" comes up. Why is it that on a geodesic path there is no turning and on a non-geodesic path there is turning? Physically, in order to avoid turning, the bug has to move its left feet the same distance as its right feet. On a non-geodesic path the bug has to walk more slowly with the legs that are on the side to which the path is turning. You can test this same idea by taking a small toy car with its wheels fixed so that, on a plane, it rolls along a straight line. Then on the surface the car will roll along a geodesic but it will not roll along other curves.

PROBLEM 1.3. Geodesics on Cones[†]

We now investigate geodesics on a cone which behave in some ways like the cylinder and in some ways differently.

a. *What lines are geodesics on (straight with respect to) the surface of a cone? Why? Have you listed all of them? How do you know?*

b. *How many geodesics join two points on a cone? Is there always at least one?*

***c.** *On a cone, can a geodesic ever intersect itself? How many times?*

If you attempt to visualize lines on a cone without looking at a paper model, you are bound to make claims that you would easily see are mistaken if you investigated them on an actual cone. **You must make models of cones**. And you must look at cones of different shapes, i.e., cones with varying cone angles (see the next page). Try the activities mentioned in the paragraph preceding Problem **1.3**.

Lay a stiff ribbon or straight strip of paper on a cone. Convince yourself that it will follow a straight line with respect to the surface. Also, convince yourself that straight lines on the cone, when looked at locally and intrinsically, have the same symmetries as on the plane. Finally, also consider line symmetries on the cone. Check to see if the symmetries you found on the plane will work on cones, and remember to

[†]Problems or Sections preceded by an asterisk (*) are not essential for later in this book.

think intrinsically and locally. A special class of geodesics on a cone are the ***generators***: the straight lines that go through the cone point. These lines have some extrinsic symmetries (can you see which ones?), but in general, geodesics have only local, intrinsic symmetries. For example, can any geodesic that is not a generator have global extrinsic reflection symmetries? Why?

Walking along a generator: When looking at straight paths on a cone, you will be forced to consider straightness at the cone point. See Figure 1.9. You might decide that there is no way the bug can go straight once it reaches the cone point, and thus a straight path leading up to the cone point ends there. Or you might decide that the bug can find a continuing path that has most of the symmetries of a straight line on the plane. Do you see which path this is?

Figure 1.9. What is straight through the cone point?

Geodesics behave differently on differently shaped cones. So an important variable is the cone angle. The ***cone angle*** is generally defined as the angle measured around the point of the cone on the surface. Notice that this is an intrinsic description of angle. The bug could measure a cone angle by first making a model of a one-degree angle and then, determining how many of the one-degree angles it would take to go around the cone point. If we use radian measure, then the cone angle is c/r, where c is the circumference of the circle (on the cone) which is at a distance r from the cone point. We can determine the cone angle extrinsically in the following way: If we cut the cone along a generator and flatten it, then the cone angle is the angle of the planar sector. For example, if we take a piece of paper and bend it so that half of one side meets up with the other half of the same side as in Figure1.10, we will have a 180°-cone:

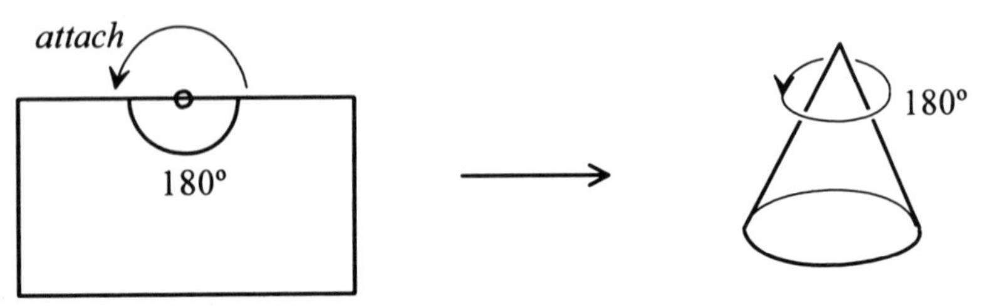

Figure 1.10. 180° cone.

A 90°-cone is also easy to make—just use the corner of a sheet of paper and bring one side around to meet with the adjacent side. Also be sure to look at larger cones. One convenient way to do this is to make a cone with a variable cone angle. Take a sheet of paper and cut (or tear) a slit from one edge to the center. (See Figure 1.11.) A rectangular sheet will work but a circular sheet is easier to picture. Note that it is not necessary that the slit be straight!

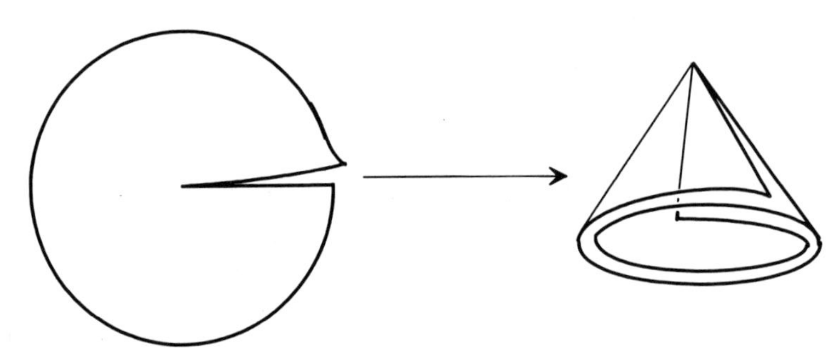

Figure 1.11. A cone with variable cone angle (0 - 360°).

You have already looked at a 360°-cone in some detail—it is just a plane. The cone angle can also be larger than 360°. A common larger cone is the 450°-cone. You probably have a cone like this somewhere on the walls, floor, and ceiling of your room. You can easily make one by cutting a slit in a piece of paper and inserting a 90° slice (360° + 90° = 450°) as pictured in Figure 1.12.

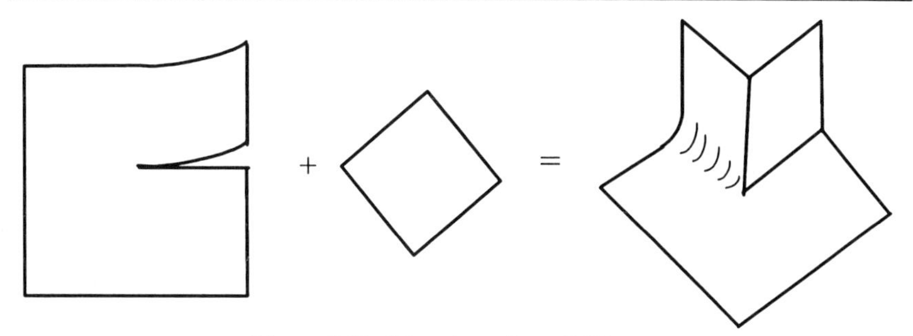

Figure 1.12. How to make a 450°-cone.

You may have trouble believing that this is a cone, but remember that just because it cannot hold ice cream, that does not mean it is not a cone. If the folds and creases bother you, they can be taken out—the cone will look ruffled instead. It is important to realize that when you change the shape of the cone like this (i.e., by ruffling), you are only changing its extrinsic appearance. Intrinsically (from the bug's point of view) there is no difference.

You can also make a cone with variable angle of more than 360° by taking two sheets of paper and slitting them together to their centers as in Figure 1.13. Then tape the left side of the top slit to the right side of the bottom slit as pictured.

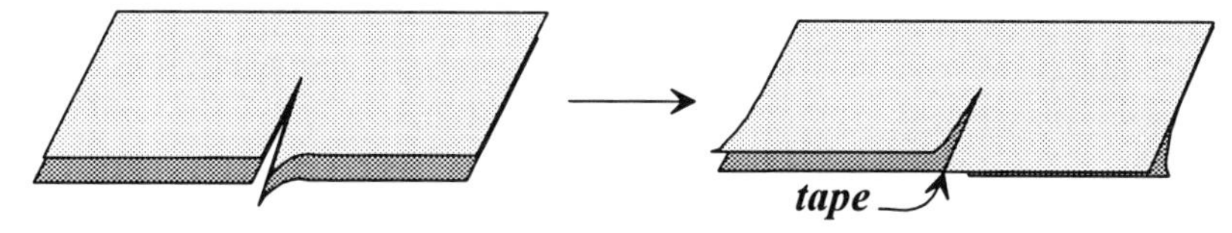

Figure 1.13. Variable cone angle larger than 360°.

It may be helpful for you to discuss some definitions of a cone. The following is one definition: *Take any simple (non-intersecting) closed curve* ***a*** *on a sphere and consider a point* ***P*** *at the center of the sphere. A* ***cone*** *is the union of the rays that start at* ***P*** *and go through each point on* ***a***. (See Figure 1.14.) The cone angle is then equal to

(length of ***a***)/(radius of sphere),

in radians. Do you see why? Experiment by making out of paper examples of cones like those shown above.

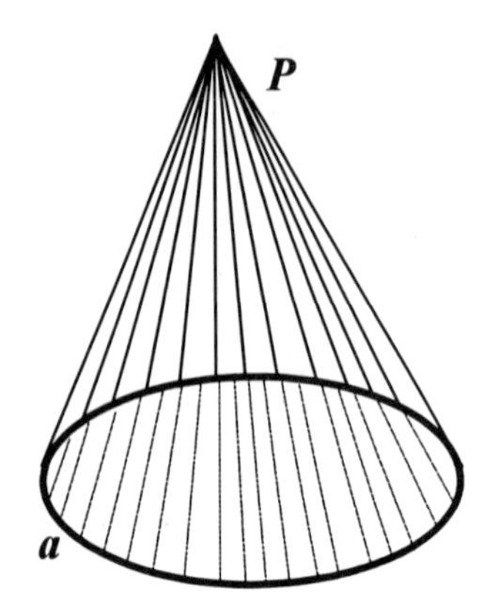

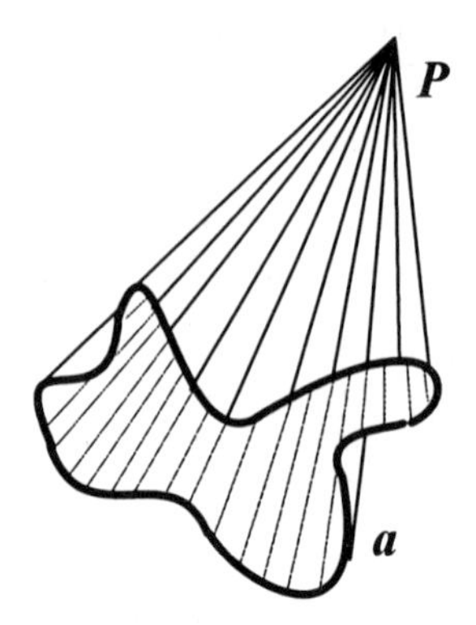

Figure 1.14. General cones.

***d.** *On a 450° cone find a point P (not the cone point) and a geodesic l (not through the cone point) such that there are many geodesics through P that do not intersect l. Compare this situation to the usual parallel postulate for the plane.*

In standard treatments of non-Euclidean geometries, hyperbolic geometry is presented as a geometry in which there are more than one line through a given point parallel to a given line. Euclid's fifth postulate for the plane implies that on the plane there is exactly one line through a given point parallel to a given line. For a discussion about parallel postulates, see Chapter 10 of the author's *Experiencing Geometry on Plane and Sphere,* [**SP**: Henderson].

Is "Shortest" Always "Straight"?

We are often told that "a straight line is the shortest distance between two points," but is this really true? As we have already seen on a cylinder, two points are, in general, connected by at least two straight paths. Only one of these paths is the shortest. The other is also straight, but not the shortest straight path.

Consider a model of a cone with angle 450°. Notice that such cones appear commonly in buildings as so-called "outside corners" (see Figure 1.15). It is best, however, for you to have a paper model that can be flattened. Use your model to investigate which points on the cone can be joined by straight lines. In particular, look at points like those labeled *A* and B in Figure 1.15. There is no single straight line on the cone going from A to B, and thus for these points the shortest path is not straight. Convince yourself that in this case this shortest path is not straight. (This is part of 1.3.b.)

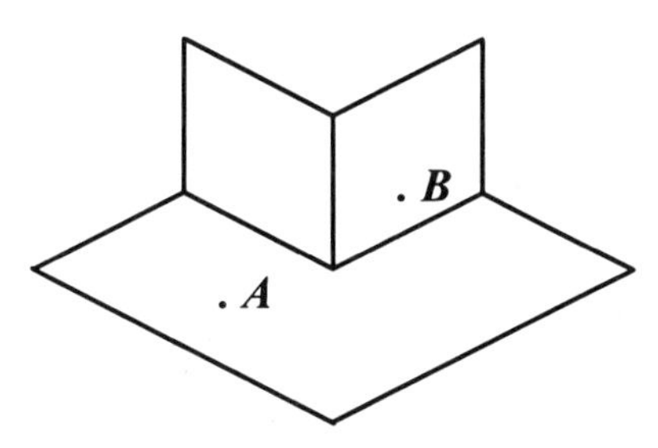

Figure 1.15. There is no geodesic path from A to B.

Here is another example: Think of a bug crawling on a plane with a tall box sitting on that plane (refer to Figure 1.16). This combination surface—the plane with the box sticking out of it—has eight cone points. The four at the top of the box have 270° cone angles, and the four at the bottom of the box have 450° cone angles (180° on the box and 270° on the plane). What is the shortest path joining points X and Y, which are on opposite sides of the box? Is the straight path the shortest? Is the shortest path straight? To check that the shortest path is not straight, see that at the bottom corners of the box, the two sides of the path have different angular measures.

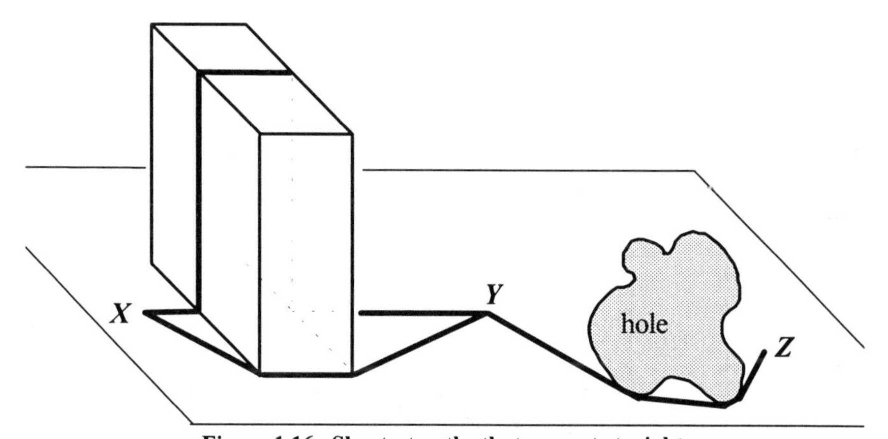

Figure 1.16. Shortest paths that are not straight.

Also consider a planar surface with a hole removed as in Figure 1.16. Check that for points *Y* and *Z*, the shortest path (on the surface) is not straight because the shortest path must go around the hole.

So, we see that sometimes a straight path is not shortest and the shortest path is not straight. However, for surfaces that are "smooth" enough, there is a close relationship between "straight" and "shortest." A *smooth* surface is essentially what it sounds like—a surface is smooth at a point if, when you zoom in on the point, the surface becomes indistinguishable from a flat plane. Note that a cone (with cone angle not equal to 360°) is not smooth at the cone point but is smooth at all other points; also a sphere and a cylinder are both smooth at every point. The surface of a piece of paper with a crease in it is not extrinsically smooth but it is *intrinsically smooth* in the sense that locally and intrinsically its geometry is the same as the plane. The following is a theorem which we will prove in a later chapter:

> **Theorem:** *If a surface is smooth (in the* C^2 *sense), then a geodesic on the surface is always the shortest path between "nearby" points. If the surface is also geodesically complete (that is, every geodesic on it can be extended indefinitely, for example, there are no holes), then any two points can be joined by a geodesic which is the shortest path between them.* (See Problem **7.4**.)

A surface is C^2 if it can be described by local coordinates (see the next section) whose first and second derivatives exist and are continuous. This is a stronger condition than merely being extrinsically smooth (see Problems **3.1** and **6.1.c**). However, I do not know if there are any surfaces that are intrinsically smooth for which there are is no C^2 embedding into Euclidean space or for which the theorem above is false.

We encourage the reader to discuss how each of the previous examples is in harmony with this theorem. Note that the statement "every geodesic on it can be extended indefinitely" is a reasonable interpretation of Euclid's first postulate, which says "every line can be extended indefinitely." We will begin a detailed discussion of smooth surfaces in Chapter 3.

Locally Isometric Surfaces

We can describe this situation more generally by defining: Two geometric spaces, `G` and `H`, are said to be ***locally isometric*** at points *G* in G and H in H if the local intrinsic experience at G is the same as the experience at H. That is, there are neighborhoods of G and H that are identical in terms of intrinsic geometric properties such as measurement of lengths and angles in the neighborhoods. A cylinder and the plane are locally isometric (at every point), and the plane and a cone are locally isometric except at the cone point. Two cones are locally isometric at their cone points only if the cone angles are the same.

Euclid defines a right angle as follows: "When a straight line set up on a straight line makes the adjacent angles equal to one another, each of the equal angles is **right**" [**AT**: Euclid's *Elements*]. Note that if you use this definition, then right angles at a cone point are not equal to right angles at points that are locally isometric to the plane. Euclid goes on to state as his fourth postulate: "All right angles are equal to one another." Thus, Euclid's postulate rules out cone points.

A surface that is ***locally isometric*** to the plane is traditionally called ***developable***. The notion of developable is important, for example, in the manufacture of the steel hull of ships. Those portions of the hull that are developable surfaces can be made by bending a sheet of steel; but those portions of hull that are not developable must be covered with more expensive "furnace plates," which are steel sheets that have been softened in a furnace and then molded into the desired shape.

Local Coordinates for Cylinders and Cones†

If we have a surface M in 3-space then ***local coordinates*** (or a ***local coordinate patch***) for M is a continuous function of two real variables, defined in a region R in the plane, which maps R one-to-one onto some region of M by describing the point in the region, which has coordinates, a, b. See Figure 1.17. We will use lower case boldfaced letters to denote the function that defines the local coordinates. We call these ***extrinsic local coordinates*** if the location of the point is described extrinsically (usually this means in terms of its rectangular coordinates in 3-space or $\mathbb{R}^n$) and we call them ***intrinsic local coordinates*** if the location of the point is described intrinsically by referring only to intrinsic geometric properties of the surface.

†In Chapters 1 though 5 of this text, local coordinates are not necessary for the understanding of any of the main geometric concepts except in Problem **4.7**. Thus readers who find local coordinates distracting at this stage may skip all sections and parts of problems dealing with local coordinates while reading the first five chapters.

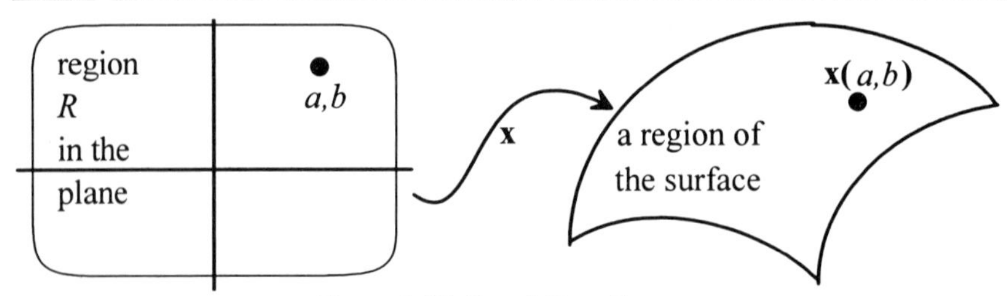

Figure 1.17. Local Coordinates.

For example, if we have a cylinder in 3-space, then to define extrinsic local coordinates, we can:

1. Choose an origin for 3-space on the axis of the cylinder;
2. Choose two straight rays from the origin, both rays perpendicular to each other and to the axis, which we call the x-axis and the y-axis;
3. Choose one of the two directions along the axis of the cylinder as the positive z-axis.

Then one possible extrinsic local coordinates are:

$$\mathbf{x}(\theta,z) = (r \cos \theta, r \sin \theta, z),$$

where r is the radius of the cylinder.

This definition would not be appropriate for a 2-dimensional bug on the surface because the bug has no awareness of 3-space. Instead the bug would like to define the local coordinates intrinsically. To do this the bug could:

1. Choose any point on the cylinder as its (intrinsic) origin, the point with coordinates, (0,0), written $\mathbf{y}(0,0)$;
2. Choose at $\mathbf{y}(0,0)$ one of two directions along the unique geodesic that comes back to $\mathbf{y}(0,0)$, which he might call the positive direction along the base curve;
3. Choose one of two geodesic rays at $\mathbf{y}(0,0)$ which are perpendicular to the base curve as the positive z-axis.

Then the intrinsic local coordinates can be described as

> $\mathbf{y}(w,z)$ = {The point attained by walking along the base curve a distance w and then turning at right angles in the direction of the positive z-axis walking along that geodesic a distance z.}

(See Figure 1.18.)

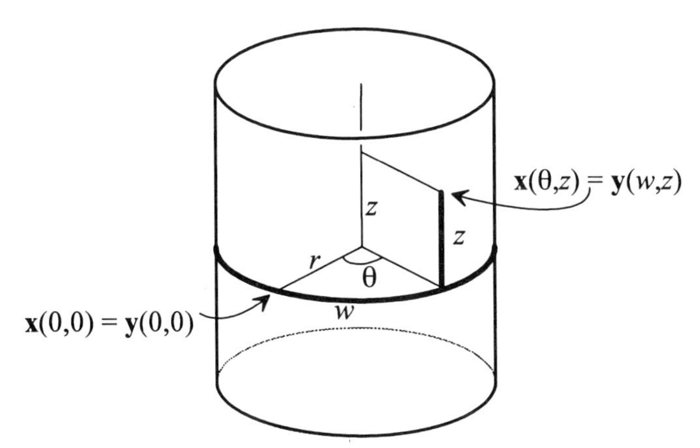

Figure 1.18. Local coordinates on a cylinder.

Such local coordinates are called ***geodesic rectangular coordinates***. In this example the base curve is a geodesic, but that is not necessary. The same intrinsic description will work with any smooth curve chosen as the base curve. However, it is necessary (as we will see later) that the curves defined by the second coordinates be geodesics.

These intrinsic and extrinsic local coordinates are different in the sense that, for most points, $\mathbf{x}(a,b) \neq \mathbf{y}(a,b)$.

For a cone, the natural origin (both intrinsically and extrinsically) is the cone point. The angle that the 2-dimensional bug would measure at the cone point is called the ***cone angle*** α. If we use radian measure, then $\alpha = c/r$, where c is the circumference of the circle on the cone, which is at a distance r from the cone point. The ***geodesic polar coordinates*** on the cone can be described intrinsically by

> $\mathbf{y}(\theta,r)$ = {the point $\mathbf{p}$ on the cone, where r is the length of the line segment from $\mathbf{p}$ to the cone point and θ is the angle along the surface between this segment and a fixed reference ray from the cone point}. (See Figure 1.19.)

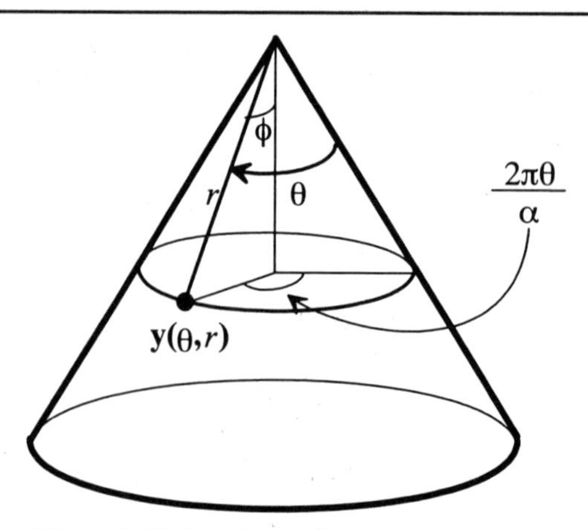

Figure 1.19. Local coordinates on a cone.

In the same manner as polar coordinates on the plane, we allow θ to be any angle, but note that two angle coordinates denote the same point on the cone if they differ by an integral multiple of α. You should convince yourself that *these intrinsic polar coordinates work for any cone, even those with cone angle larger than* 2π.

We can define ***geodesic polar coordinates*** extrinsically on a right circular cone with cone angle α by

$$\mathbf{x}(\theta, r) = (r \sin\phi \cos\frac{2\pi\theta}{\alpha}, r\sin\phi\sin\frac{2\pi\theta}{\alpha}, r\cos\phi),$$

where ϕ is the angle between the axis of the cone and a generator of the cone. By looking at a circle of radius 1 from the cone point the reader can check that its circumference is $\alpha = 2\pi \sin \phi$. Note that *the extrinsic definition only works for cones with cone angle less than* 2π.

Other local rectangular coordinates can be seen by placing graph paper in various orientations on a cone or cylinder.

Problem 1.4. Geodesics in Local Coordinates

***a.** *On the cylinder give an intrinsic definition of the coordinate patch* **x**, *which is defined extrinsically above. Also give an extrinsic definition of the coordinate patch* **y** *which is defined intrinsically above.*

b. *Intrinsically define on a cone* **geodesic rectangular coordinates** *with the base curve being one of the circles at a fixed distance from the cone point. Compare these coordinates with geodesic polar coordinates on the same cone.*

c. *If γ is a geodesic on the cylinder and α is the angle γ makes with the base curve, then show that in terms of intrinsic rectangular coordinates the parametric equations for γ are*

$$\gamma(s) = \mathbf{y}(s \cos \alpha, s \sin \alpha),$$

where s is the arclength along γ from the (intrinsic) origin. Write an equation for γ in terms of extrinsic local coordinates. Given two points p and q on the cylinder, determine which geodesics join p to q.

d. *In terms of intrinsic polar coordinates, show that if λ is a geodesic on the cone and $\mathbf{p} = (d,\beta)$ is the point on the geodesic that is closest to the cone point, then an arbitrary point $\mathbf{y}(\theta,r)$ on the geodesic satisfies the equation*

$$r = d \sec(\theta - \beta).$$

Show that the geodesic can be parametrized by arclength s as follows:

$$r = \sqrt{d^2 + s^2}\,,\ \theta = \beta + \arctan \frac{s}{d}.$$

***e.** *Write an equation for a geodesic on a cone in terms of extrinsic local coordinates.*

***f.** *How many times does a geodesic on the cone intersect itself? How does the number of self-intersections depend on the cone angle?*

Problem 1.5. What Is Straight on a Sphere?

a. *Imagine yourself to be a bug crawling around on a sphere. (This bug can neither fly nor burrow into the sphere.) The bug's universe is just the surface; it never leaves it. What is "straight" for this bug? What will the bug see or experience as straight? How can you convince yourself of this? Use the properties of straightness (such as symmetries), which you talked about in Problem 1.1., to show that the great circles are straight with respect to the sphere.*

b. *Show* (that is, convince yourself, and give an argument to convince others) *that no other circles* (for example, latitude circles) *on the sphere have the same local symmetries as the great circles.*

In Chapter 3 we will show in another way that the great circles are the only geodesics on the sphere.

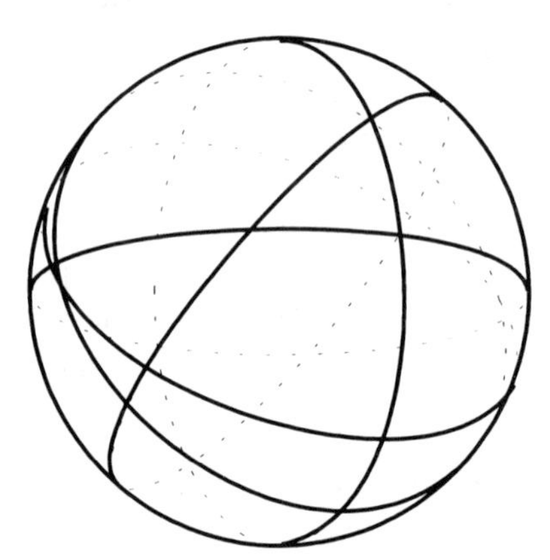

Figure 1.20. Great circles.

Great circles are those circles that are the intersection of the sphere with a plane through the center of the sphere. Examples: Any longitude line and the equator are great circles on the earth. Consider any pair of opposite points as being the poles, and thus the equator and longitudes with respect to any pair of opposite points will be great circles. (See examples illustrated in Figure 1.20.)

Suggestions

The first step to understanding this problem is to convince yourself that great circles are straight lines on a sphere. Think what it is about the great circles that would make the bug experience them as straight. To better visualize what is happening on a sphere (or any other surface, for that matter), **you must use models**. This is a point we cannot stress enough. You must make lines on a sphere to fully understand what is straight and why. An orange or an old, worn tennis ball works well as a sphere, and rubber bands make good lines. Also, you can use ribbon or strips of paper. Try placing these items on the sphere along different curves to see what happens.

Also look at the symmetries from Problem 1.1 to see if they hold for straight lines on the sphere. The important thing to remember here is to **think in terms of the surface of the sphere, not in 3-space**. Always try to imagine how things would look from the bug's point of view.

Experimentation with models plays an important role here. Working with models *that you create* helps you to experience that great circles are, in fact, the only straight lines on the surface of a sphere. Convincing yourself of this notion will involve recognizing that straightness on the plane and straightness on a sphere have common elements. The activities listed at the end of Problem **1.3** are all usefull here also. However, it is better for you to come up with your own experiences.

These activities will provide you with an opportunity to investigate the relationships between a sphere and the geodesics of that sphere. Your experiences should help you to discover how the symmetries of great circles are mostly the same as the symmetries of straight lines on a plane.

Also notice that, on a sphere, straight lines are circles (points on the surface a fixed distance away from a given point) — special circles whose circumferences are straight! Note that the equator is a circle with two centers: the north pole and the south pole. In fact, any circle on a sphere has two centers.

Intrinsic Curvature on a Sphere

You have tried wrapping the sphere with a ribbon and noticed that the ribbon will only lie flat along a great circle. (If you haven't experienced this yet, then do it now before you go on.) Arcs of great circles are the only paths of a sphere's surface that are tangent to a straight line on a piece of paper wrapped around the sphere. If you wrap a piece of paper tangent to the sphere around a latitude circle (see Figure 1.21), then, extrinsically, the paper will form a portion of a cone and the curve on the paper will be an arc of a circle. The *intrinsic* (or *geodesic*) *curvature* of a path on the surface of a sphere can be defined as the curvature that one obtains when one "unwraps" the path onto a plane. For more details on intrinsic curvature see Chapter 3.

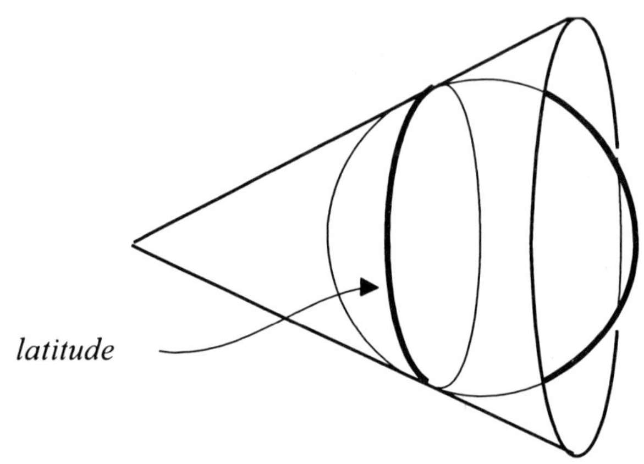

Figure 1.21. Finding the intrinsic curvature.

In Chapter 3 we will talk about geodesics in terms of the velocity vector of the motion as one travels at a constant speed along that path. (The velocity vector is tangent to the curve along which the bug walks.) For example, as you walk along a great circle, the velocity vector to the circle changes direction, extrinsically, in 3-space where the change in direction is toward the center of the sphere. "Toward the center" is not a direction that makes sense to a 2-dimensional bug whose whole universe is the surface of the sphere. Thus, the bug does not experience the velocity vectors at points along the great circle as changing direction. In Chapter 8, the rate of change, from the bug's point of view, is called the *covariant* (or *intrinsic*) *derivative*. We will show that as the bug traverses a geodesic, the covariant derivative of the velocity vector is zero. Geodesics can also be expressed in terms of *parallel transport*, discussed in Chapter 5.

Local Coordinates on a Sphere

Extrinsically we can express the sphere of radius R with center at the origin of 3-space by

$$\mathbf{x}(\theta,\phi) = (R \cos \phi \cos \theta, R \cos \phi \sin \theta, R \sin \phi).$$

(See Figure 1.22.)

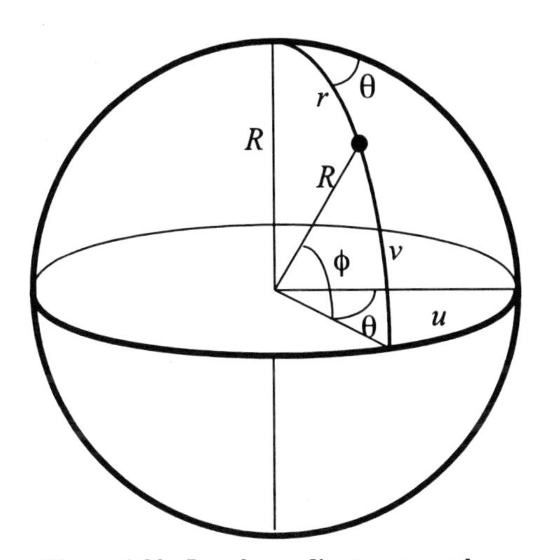

Figure 1.22. Local coordinates on a sphere.

This expression provides local coordinates for the sphere except at the North and South Poles ($\phi = \pm\pi/2$). Except when $R = 1$, $\mathbf{x}$ is not strictly geodesic rectangular coordinates because, even though the longitudes (θ equal to a constant) are arcs of great circles, the coordinate ϕ does not express the arclength as is required by geodesic coordinates. The reader should check that the following is an (extrinsic) expression of geodesic rectangular coordinates on the sphere of radius R:

$$\mathbf{y}(u,v) = (R \cos(v/R) \cos(u/R), R \cos(v/R) \sin(u/R), R \sin(v/R)).$$

Geodesic Polar Coordinates at the poles can be expressed as

$$\mathbf{z}(\theta,r) = (R \sin(r/R) \cos\theta, R \sin(r/R) \sin\theta, R \cos(r/R)).$$

PROBLEM 1.6. Strakes, Augers, and Helicoids

To give structural support to large metal cylinders, such as large smoke stacks, engineers sometimes attach a spiraling strip called a *strake.*† (See Figure 1.23.) The strake and related surfaces are common surfaces in the physical world and will serve as illustrative examples throughout this text.

To produce the strake it is convenient to cut annular pieces (the region between two concentric circles) from a flat sheet of steel as illustrated in the Figure 1.23.

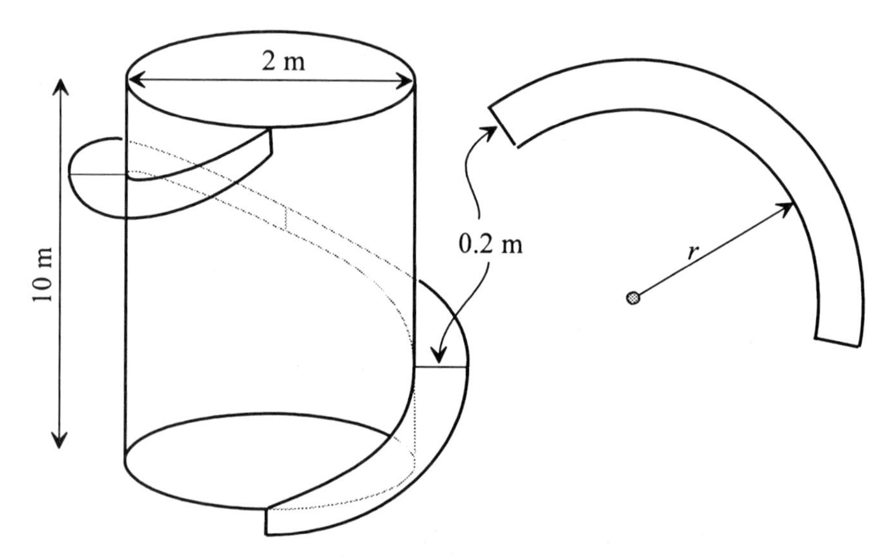

Figure 1.23. A strake.

These annular pieces are then bent along a helix on the cylinder to form the strake. In Chapter 2 we will convince ourselves that the way to compute the ideal value for the radius r is to require that the helix on the cylinder and the inner curve of the annulus to have the same curvature. Also in Chapter 2 we will ask if the flat annulus can be bent (but not stretched) to fit the strake so that the radial line segments of the strake are horizontal straight lines. In other words: *Is the strake developable?*

What happens if we make the strake very wide compared to the diameter of the cylinder—such as happens in an auger? (See Figure 1.24.)

†This example is inspired by an example given in [**DG**: Morgan].

If we double the auger (extend it in both directions from the cylinder) and then shrink the cylinder to zero radius, the resulting figure is called a ***helicoid***.

We suggest that at this point you use Computer Exercise 1.6 in Appendix C to display computer images of the strake/auger/helicoid with a cylinder of variable radius.

a. *Show that the horizontal line segments in the strake and helicoid have intrinsic mirror symmetry and thus must be geodesics on the strake or helicoid.*

[Hint: Look at the extrinsic 180° rotation about the segment.]

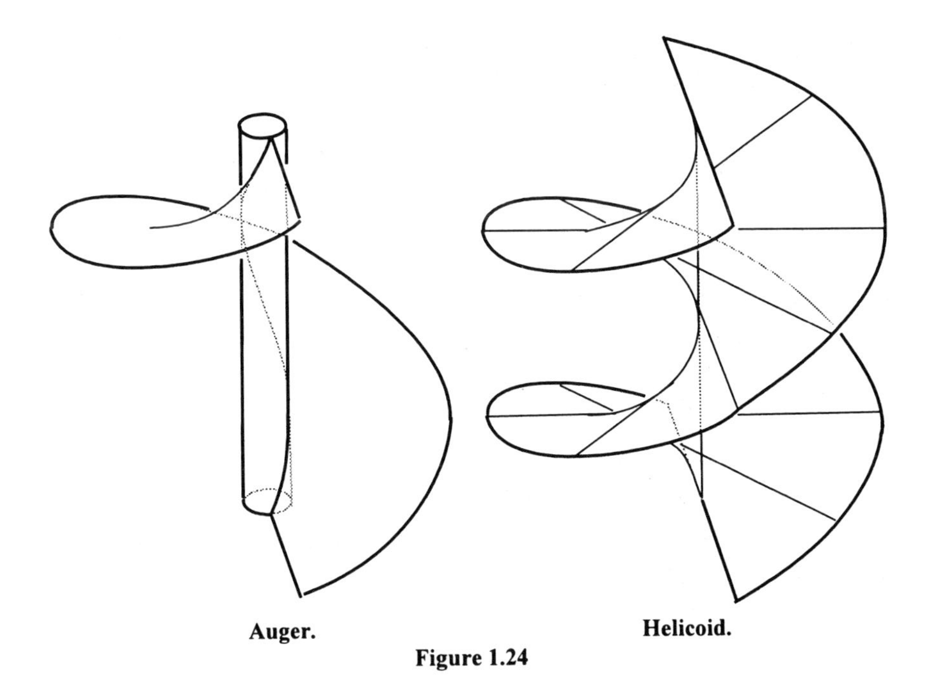

Figure 1.24

b. *Show that, if h is the measure of how high the helix goes on the cylinder in one turn of the helix, then arclength along one turn of the helix is*

$$\sqrt{h^2 + (2\pi R)^2}$$

[Hint: Look at the helix from the bug's 2-dimensional point-of-view.]

c. *Find an extrinsic expression for the helix, parametrized by arclength. (That is, express the helix as*

$$\gamma(s) = (x(s), y(s), z(s)),$$

where s is arclength.)

d. *Find an extrinsic expression for geodesic rectangular coordinates on the strake with the helix along the cylinder as the base curve.*

[Hint: Start with Part **c**.]

***e.** *How do you need to modify the coordinates in part* **d** *so that they will be local coordinates for a helicoid with the center line of the helicoid as the base curve?*

[Hint: What happens as $R \to 0$?]

Problem 1.7. Surfaces of Revolution

If f is a positive-valued function of one real variable, then

$$\mathbf{x}(\theta,x) = (x, f(x)\cos\theta,\ f(x)\sin\theta)$$

is extrinsic coordinates defining a surface of revolution (revolved about the x-axis). (See Figure 1.25.)

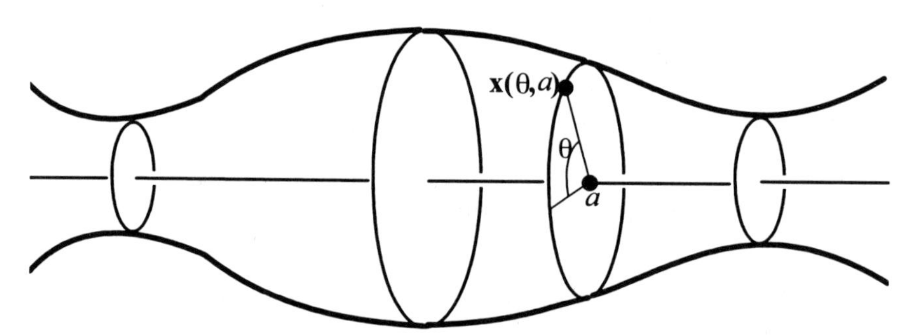

Figure 1.25. Surface of revolution.

Computer Exercise 1.7 will allow you to use the computer to display surfaces of revolution. Can you display a cylinder? a cone? a sphere? a paraboloid? Keep these surfaces in mind as you do Problem 1.7.

a. *Argue that the curves on the surface of revolution that have constant* θ *are geodesics.*

b. *Which generating circles* (x = constant) *appear to be geodesics? Why?* We will return to this question in Chapters 3 and 6.

[Hint: Imagine laying a ribbon on the surface.]

c. *On a surface of revolution describe geodesic rectangular coordinates with one of the generating circles* (x = constant) *as base curve.*

[Hint: See pages 20-22 for a description of geodesic rectangular coordinates. Remember that the base curve and the second coordinate curves (geodesic) must be parametrized by arclength.]

PROBLEM 1.8. Hyperbolic Plane

Hyperbolic geometry, discovered more than 150 years ago by C.F. Gauss (German), J. Bolyai (Hungarian), and N.I. Lobatchevsky (Russian), is special from a formal axiomatic point of view because it satisfies all the axioms of Euclidean geometry except for the parallel postulate. In hyperbolic geometry there are many straight lines through a given point that do not intersect a given line. (Compare with Problem **1.3.d**.)

Hyperbolic geometry and non-Euclidean geometry are often considered as being synonymous, but as we have seen there are many non-Euclidean geometries, particularly spherical geometry. It is also not accurate to say (as many books do) that non-Euclidean geometry was discovered more than 150 years ago. Spherical geometry (which is clearly not Euclidean) was in existence and studied by at least the ancient Babylonians, Indians, and Greeks more than 2,000 years ago. Spherical geometry was of importance for astronomical observations and astrological calculations. Even Euclid in his *Phaenomena* [**AT**: Euclid] (a work on astronomy) discusses propositions of spherical geometry. Menelaus, a Greek of the first century, published a book *Sphaerica*, which contains many theorems about spherical triangles and compares them to triangles on the Euclidean plane. (*Sphaerica* survives only in an Arabic version. For a discussion see [**NE**: Kline, page 120].)

A paper model of the hyperbolic plane may be constructed as follows[†]: Cut out many identical annular ("annulus" is the region between two concentric circles) strips as in the following Figure 1.26. Attach the strips together by attaching the inner circle of one to the outer circle of the other or the straight ends together. The resulting surface is of course only an approximation of the desired surface. The actual hyperbolic plane is obtained by letting $\delta \to 0$ while holding r fixed. Note that since the surface is constructed the same everywhere (as $\delta \to 0$), it is ***homogeneous*** (i.e. intrinsically and geometrically, every point has a neighborhood that is isometric to a neighborhood of any other point). We will call the results of this construction the ***annular hyperbolic plane.*** In Parts **c** and **d**, below, there is a different description of the hyperbolic plane.

I strongly suggest that the reader at this point take the time to cut out carefully several such annuli and to tape them together as indicated.

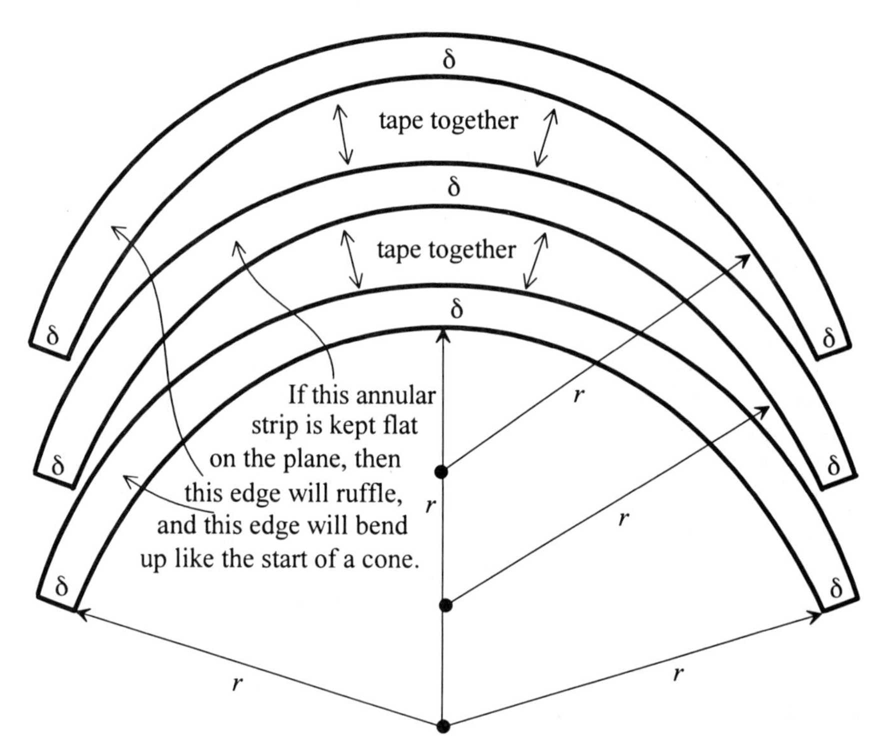

Figure 1.26. Annular strips for making an annular hyperbolic plane.

[†]The idea for this construction was shown to me by William Thurston in 1978.

a. *Argue that the curves on the annular hyperbolic plane, which run radially across each annular strip, are geodesics.*

[Hint: Look for the local intrinsic symmetries.]

b. *On the annular hyperbolic plane describe geodesic rectangular coordinates with the base curve being one of the circular edges of one of the annuli and the positive second coordinate direction being along the geodesics that run radially across each annular strip toward the center of the annulus.*

[Note that it only makes sense now to do this intrinsically.]

c. *Let λ and μ be two of the geodesics described in part* **a**. *If the distance between λ and μ along the base curve is d, then show that the distance between them at a distance $c = n\delta$ from the base curve is on the paper hyperbolic model:*

$$d\left(\frac{r}{r+\delta}\right)^{n} = d\left(\frac{r}{r+\delta}\right)^{c/\delta}.$$

Now take the limit as $\delta \to 0$ to show that the distance between λ and μ on the annular hyperbolic plane is

$$d \exp(-c/r).$$

Now we define a new coordinate patch **z** from the upper half plane

$$\mathrm{H}^{+} \equiv \{(x,y) \in \mathbb{R}^2 \mid y > 0\}$$

onto the annular hyperbolic plane with $r = 1$ by defining

$$\mathbf{z}(x,y) = \mathbf{x}(x,\ln(y)),$$

where **x** is the geodesic rectangular coordinates defined in Part **b**. This representation of the hyperbolic plane is usually called ***the upper half plane model***.

The particular form of these coordinates is such that **z** is ***conformal***, meaning that as **z** takes the upper half plane onto the hyperbolic plane, it does not change angles (but will distort distances). To see that **z** is conformal, note first that the x and y coordinate curves in the plane and on the annular hyperbolic plane are both orthogonal to each other. Other angles will not be changed if, at each point, the distortion of distances along the two coordinate curves is the same, because then the distortion at each point will be infinitesimally a similarity (which preserves angles). For any curve γ the ***distortion*** at a point $p = \gamma(a)$ is

$$\lim_{x \to a} \frac{\text{the arc length along } \gamma \text{ from } \gamma(a) \text{ to } \gamma(x)}{|x - a|}.$$

If the curve lies in Euclidean space, then this is precisely the speed $|\gamma'(a)|$. Thus we need to:

d. *Show that the distortion of both coordinate curves*

$$x \to \mathbf{z}(x,b) \quad \textit{and} \quad y \to \mathbf{z}(a,y)$$

at the point $\mathbf{z}(a,b)$ *is* $1/b$.

[Hint: For the first coordinate direction, use the result of Part **c**. For the second coordinate direction, use the fact that the second coordinate curves in geodesic rectangular coordinates are parametrized by arclength.]

In Problem **7.5** you will show that there exist in the hyperbolic plane: rotations through any angle about any point, reflections through any geodesic, and translations along any geodesic segment.

It is a theorem (see Problem **3.1**) that there does not exist a smooth (C^1) isometric embedding of the whole hyperbolic plane into 3-space. However, the paper model can be extended indefinitely and provides (as $\delta \to 0$) an isometric embedding in 3-space. For a small section of the surface it is possible to have the embedding smooth (see Problem **3.1**), but it is possible to see in the model that as you take larger and larger sections, eventually "ruffles" will form, which cause the surface not to have a tangent plane at some points and thus not to be smooth.

Problem 1.9. Surface as Graph of a Function $z = f(x,y)$

If we have any real-valued function $z = f(x,y)$ defined on a region in the plane, then we can use it to define coordinates extrinsically for a surface

$$\mathbf{x}(x,y) = (\, x, y, f(x,y)\,).$$

On the other hand, if M is a smooth surface in 3-space, which has a tangent plane T_pM at the point p, then we may find a function whose graph is a neighborhood of p in the surface. Choose coordinates (x,y,z) in 3-space so that $p = (0,0,0)$ and the tangent plane T_pM is the (x,y)-plane. The projection

$$g(x,y,z) = (x,y,0)$$

will be one-to-one onto a neighborhood of p. (See Appendix B.) Thus there will be an inverse function $\mathbf{x}(x,y) \in M$ of the form

$$\mathbf{x}(x,y) = (x,y,f(x,y)).$$

Therefore $\mathbf{x}$ expresses the portion of M in a neighborhood of p as the graph of a function. These coordinates have the extra property that the plane tangent to M at p is the (x,y)-plane and $p = (0,0,0)$. Such special graphs of functions are called a ***Monge patch***. (See Figure 1.27.) In general it is not possible (or is very difficult), given a surface, to explicitly find the function that defines a Monge patch. We will show in Problem **3.1.e** that such Monge patches for a smooth manifold will be continuously differentiable.

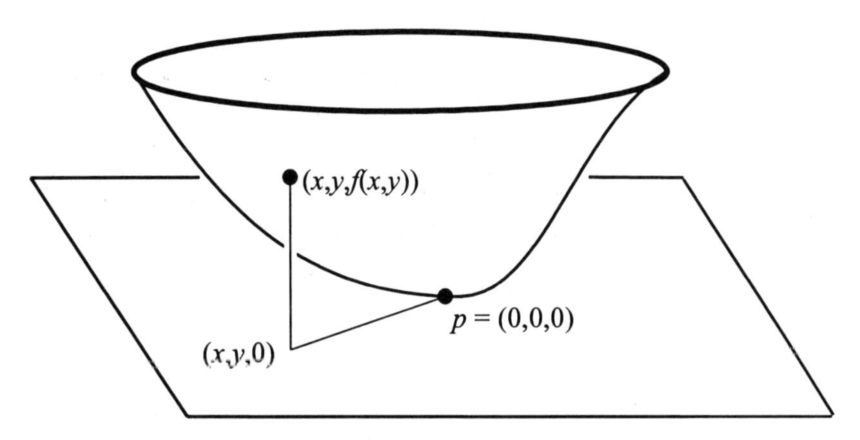

Figure 1.27. Monge patch.

You are encouraged to use Computer Exercise 1.9 to display the graph of any real-valued function of two real variables.

a. *Express locally as a graph of a function a helicoid and a cone with angle less than* 360°.

b. *Find a Monge patch for a cylinder and for a sphere.*

c. *Consider a general surface of revolution*

$$\mathbf{x}(\theta,x) = (x, f(x)\cos\theta,\ f(x)\sin\theta).$$

Express the surface locally as the graph of a function. If the derivative $f'(a) = 0$*, then find a Monge patch for a neighborhood of*

$$(a, f(a)\cos\theta, f(a)\sin\theta).$$

Chapter 2
Extrinsic Curves

Introduction

The starting point of our extrinsic investigations is *views of space*. As is our ordinary experience of space, only a certain bounded region of space is within our ***field of view*** (***f.o.v.***) and within this field of view there are details too small to be ***distinguished***. This makes sense in our experience of physical space, in computer graphics images, in fixed-point and floating point arithmetic, and also applies to the spaces of our imagination.

Borrowing from the idea of zooms in photography and computer graphics, when we want to investigate more detail of a figure we may ***zoom in on a point***. Then less of the extent of space is included in our field of view but more details are now distinguishable.

We call a figure in our f.o.v. a ***point*** if it does not have two parts which are distinguishable from each other. [Note the connection between this notion and Euclid's definition "A point is that which has no parts".] We say that two figures in the f.o.v. are indistinguishable if each point of the first figure is indistinguishable from some point of the second figure, and each point of the second figure is indistinguishable from some point of the first figure.

For simplicity we shall assume that we see all parts of this field of view with equal clarity. (That is, we ignore the so-called peripheral vision, which is a region at the edge of our field of view where we can see less detail than in the center of the field of view.) Two distinguishable points in a field of view determine a line segment. In general we can subdivide this segment into 2 parts, 3 parts, 4 parts, et cetera, until each part becomes so small that it is indistinguishable from a point. We can quantify the ***tolerance*** of our vision in a field of view as the ratio $\tau > 0$ such that (see Figure 2.1):

- Every segment (in the field of view) is ***indistinguishable from a point*** if it has length less than $\tau\rho$, where ρ is the radius of the f.o.v.
- Every segment (in the field of view) is ***distinguishable from a point*** if it has length greater than $2\tau\rho$, where ρ is the radius of the f.o.v.

This indeterminate range between $\tau\rho$ and $2\tau\rho$ is convenient because in many situations the border line between distinguishable and indistinguishable is fuzzy. Also, in pixel graphics, we need to take into account the fact that the centers of square pixels are further from the adjacent pixels on the diagonal than from the pixels that are adjacent vertically or horizontally.

We can define ***limits in a f.o.v.*** by asserting that the sequence $\{x_n\}$ converges to y if eventually x_n is indistinguishable from y.

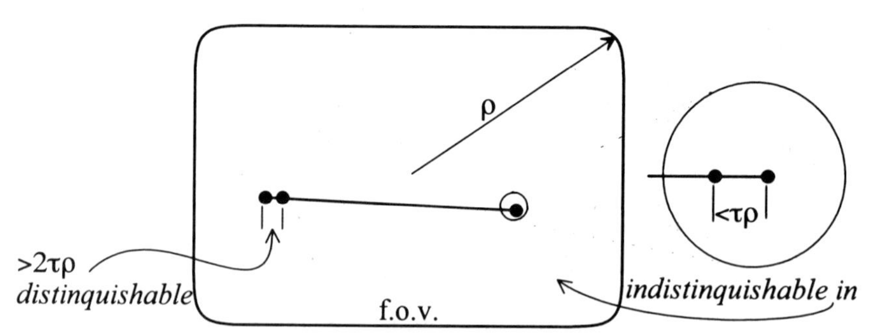

Figure 2.1. Tolerance τ and radius ρ in a f.o.v.

Problem 2.1. Give Examples of F.O.V.'s

a. *Find examples of f.o.v.'s from your experiences. Use them to illustrate the ideas in the* ***Introduction*** *section above.*

b. *Can you describe a f.o.v. that contains all three of* \$0, \$1, *and the national debt as distinguishable points? Can the quantity*

$$\{[\text{national debt}] + \$1\}$$

be distinguished from

[national debt]*?*

Answer the same questions with "$1" *replaced by* "a grain of sand" *and* "national debt" *replaced by a* "truck load of sand."

c. *Look at the real numbers in the f.o.v. of 4-digit arithmetic (fixed point).* That is, if r is a real number, then in the f.o.v. we see it as $[r]$ which is either a 4-digit non-negative integer or ∞. For example,

$$[0.2] = 0, [1.5] = 2, [2.499999] = 2,$$
$$[2.500001] = 3, [9999] = 9999, [10{,}001] = \infty.$$

If $\{p_n\}$ *is a sequence of reals converging to* q*, then can we be sure that eventually* $[p_n]$ *is equal to* $[q]$*? What about* $[p_n - q]$*? How should we define the tolerance in this f.o.v. so that*

$$\{x_n\} \to y \text{ implies that } \{[x_n]\} \to [y]?$$

Warning: It is not in general true that $p + q = r$ implies $[p] + [q] = [r]$.

***d.**† *How does* **c** *change if we use the f.o.v. of 4-digit floating point arithmetic?*

***e.** *It is meaningful for me to say "my house is one kilometer from White Hall." Is that the same as saying "my house is 1000 meters from White Hall"? Explain.*

For a discussion of related issues that come up in real number computations, see Peter R. Turner, "Will the 'Real' Real Arithmetic Please Stand Up?" [**RN**: Turner] and the book [**RN**: Moore] which exposits *interval analysis*, "an approach to computing that treats an interval as a new kind of number."

†Problems and Sections preceded by an asterisk (*) are not essential later in this book.

Archimedian Property

We further assume that the possible f.o.v.'s are restricted so that if v and u are any two f.o.v.'s then there is a third f.o.v. w in which both v and u are finite zooms; that is, the diameters and tolerances of both v and u are finitely related to the diameter and tolerance of w. (Two lengths, or ratios, $a < b$ are ***finitely related*** if there is a positive integer N such that

$$Na \leq b < (N+1)a \quad \text{or} \quad b/(N+1) < a \leq b/N.$$

This property can be called the ***Archimedian Property***, and we call the space with its f.o.v.'s an ***Archimedian space***. The Archimedian Property effectively rules out infinitesimal and infinite lengths and ratios. The standard real numbers usually studied in mathematics classes satisfy this property. However there are several "non-standard reals" that do not. See, for example, [**RN**: Conway]; [**RN**: Laugwitz]; and [**RN**: Simpson].

In an Archimedian space we can define the notion of a limit:

> **Definition:** If y and x_n ($n = 1,2,...$) are in a f.o.v. (not necessarily distinguishable–this is merely the requirement that y and the x_n are not infinite), then
>
> $$\lim_{n \to \infty} x_n = y$$
>
> if x_n is eventually indistinguishable from y in every f.o.v. containing y.

This definition is equivalent to the standard analytic definition as the interested reader can check.

Vectors and Affine Linear Space

It is assumed, for now, that our discussions take place within an Archimedian ***Euclidean space*** wherein any two distinguishable points, **p** and **q**, determine a unique straight line segment. These straight line segments (we often drop the word "straight") are distinguished geometrically by local symmetries in their neighborhoods as discussed in Problem **1.1**. If **p** is any point, then the collection of directed line segments with **p** as the initial end point are the vectors of a ***vector space*** with **p** as its origin, which we call the ***tangent space at* p**, $T_{\mathbf{p}}$. When the linear algebraic properties of this Euclidean space are being emphasized, it is usually called an ***geometric affine space***. For a detailed discussion of affine linear spaces see [**LA**: Dodson] and the *Appendix A: Linear Algebra from a*

Geometric Point of View. When we pick one point as the origin **O**, then the tangent space T_0 is usually considered as the vector space $\mathbb{R}^n$, which is the domain of normal linear algebra.

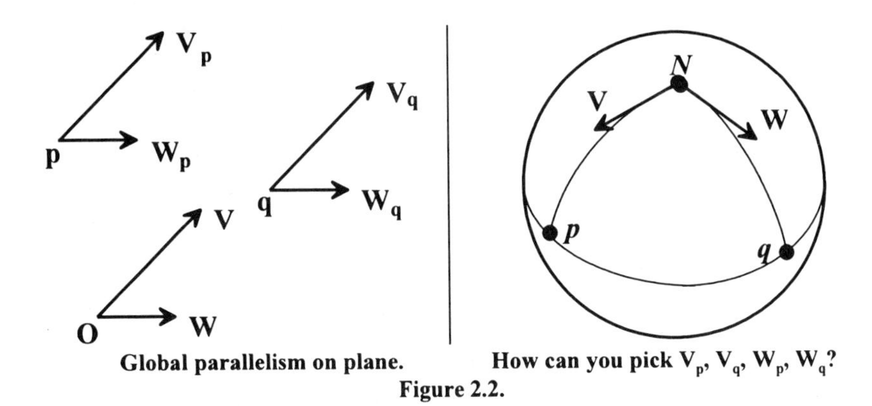

Global parallelism on plane. **How can you pick V_p, V_q, W_p, W_q?**

Figure 2.2.

In our Euclidean space there is a global notion of parallelism, so we are able to make the following definition:

> If **V** is a vector in $\mathbb{R}^n$ $(= T_0)$, then we denote by $\mathbf{V}_p$ the unique vector in T_p which is parallel to **V** (two vectors are said to be parallel if they have the same length and direction). Often, we do not even write the subscript and just call them both **V**. (See Figure 2.2.)

It is precisely this global notion of parallelism that is missing on almost all surfaces, except for a few so-called "parallelizable" surfaces, such as a cylinder. To get an experience of this phenomenon the reader should look at the North Pole and two points 90° apart on the equator. However, in Chapter 5 we will define the notion of ***parallel transport along a curve***, which makes sense on (almost) any surface.

PROBLEM 2.2. *Smoothness and Tangent Directions*

An ***infinitesimally straight curve*** in a Euclidean space is a subset in which for every point **p** on the curve there is a tangent line T_p such that, for every tolerance if you zoom in on **p**, the (orthogonal) projection of the curve onto the tangent line moves points indistinguishably. We shall call it a ***smooth curve*** if the amount of zooming necessary is *uniform* in the sense that, for each tolerance τ, there is a radius ρ such that, if we center a f.o.v. of radius less than ρ and tolerance τ at any point **p** on the curve, then within that f.o.v. the projection of the curve to the tangent line at **p** moves points indistinguishably.

It is possible for a smooth curve to have high enough curvature at **p** so that it does not look smooth at **p** in a particular f.o.v. Conversely, a physical curve may *look smooth* in a f.o.v. but is never ***infinitesimally straight*** because if you zoom in enough the smoothness will disappear.

a. *Look at the curves that are the graphs of the following functions:*

$$\sqrt{10^{-30}+x^2}\ ;\quad \frac{x}{\sqrt{10^{-30}+x^2}}\ ;\quad |x|\ ;\quad x+10^{-15}\left(\frac{x}{|x|}\right).$$

At which points are they indistinguishable from a straight line in the f.o.v. with tolerance 10^{-4} *and bounds* $-1 < x,y < 1$ *(or in the f.o.v. of a computer graphing program)? What about in other f.o.v.'s?*

[Hint: Actually view the graphs of these functions on a computer graphing program such as *Analyzer**©.]

We are thinking of a curve as a geometric object. However, it is sometimes useful to study the geometric curve by representing the curve analytically. There are two main ways of analytically representing a curve—as the graph of a function or as a parametrized curve. In first-year calculus a curve in the x-y plane is normally represented (either implicitly or explicitly) as ***the graph of a function*** $y = f(x)$. Then the derivative $f'(a)$ gives the slope of the line tangent to the graph at the point $(a, f(a))$.

b. *Prove that if the function f is differentiable at p, then the graph* $(x, f(x))$ *is infinitesimally straight at the point* $x = p$.

[Hint: If

$$t(x) = f(\mathrm{p}) + f'(p)(x-p)$$

is the equation of the line tangent to the curve $(x, f(x))$ at the point $(p, f(p))$, then

$$f(x) - t(x) = \{\ [f(x) - f(p)]/(x-p) - f'(p)\ \}(x-p).$$

The curve is infinitesimally straight if, for given tolerance ε, there is a

$$\delta > 0 \text{ (the radius of the zoom window)}$$

such that, for

$$|x - p| < \delta \text{ (} x \text{ within the zoom window),}$$

it is true that

$$|f(x) - t(x)| < \varepsilon\delta \text{ (} f(x) \text{ is indistinguishable from } t(x)\text{).}$$

In general, the value of δ might depend on p as well as on ε.]

c. *Prove that if the graph $(x, f(x))$ is infinitesimally straight at the point $x = p$ with non-vertical tangent line, then the function f is differentiable at p. Why is the restriction on the tangent line necessary?*

[Hint: Express the tangent line as

$$t(x) = f(\text{p}) + r(x - p)$$

(r is a real number) and pick a f.o.v. with radius equal to $x - p$.]

We now investigate the geometric difference between differentiable (infinitesimally straight) and continuously differentiable.

***d.** *Show that the function*

$$y = x^2 \sin(1/x)$$

is differentiable everywhere but that the derivative is not continuous at the origin. Also, show that its graph is a curve that is infinitesimally straight but not uniformly so in every neighborhood of 0.

[Hint: Investigate this graph with the help of some function graphing program, such as *Analyzer**©.]

Part **d** suggests the following result:

***e.** *Prove that a function is continuously differentiable (on closed finite intervals) if and only if its graph has no* *vertical tangents and is smooth in the sense that, for each tolerance* $\varepsilon > 0$*, there is one* $\delta > 0$ *that works for all p.*

We can also show as a corollary that

A smooth curve can always locally be written as a graph of a function.

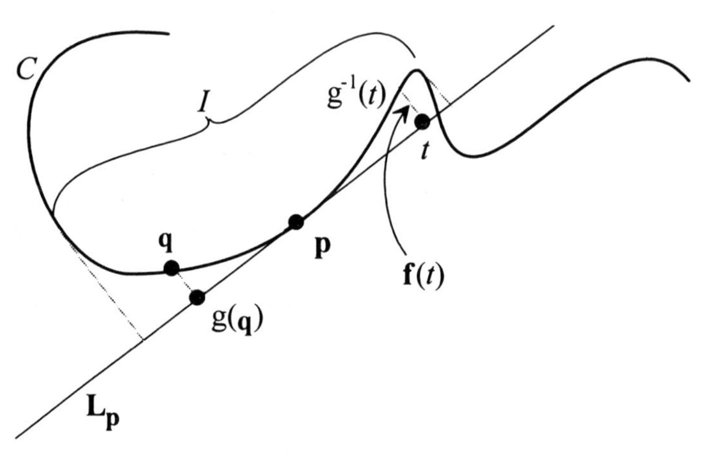

Figure 2.3. Smooth curve as graph of a function.

We now give a proof of this fact in *n*-space and picture it in the plane:

Let *C* be a smooth curve in the plane and let **p** be a point (not an end point) on *C*. Let *g* be the function that assigns to each point on *C* the orthogonal projection of that point onto $\mathbf{L_p}$, the line tangent to *C* at **p**. (A line is said to be tangent to a curve at a point if it contains the tangent vector at that point.) Then, in any interval $I \subset C$ around **p** in which the tangent directions are not perpendicular to the tangent direction at **p**, the function *g* is one-to-one and thus has an inverse g^{-1}. Then we can define $\mathbf{f}(t)$ to be the vector from *t* to $g^{-1}(t)$. Then *I* is the graph of **f** if we coordinatize *n*-space with $\mathbf{L_p}$ as the first coordinate axis and with the other axes being perpendicular to $\mathbf{L_p}$. (See Figure 2.3.) It follows from **2.2.c-d** that **f** is continuously differentiable.

For geometric investigations it is often best to ***parametrize*** the curve. Think of a particle (or bug) moving along the curve; then a

parametrization, $\mathbf{p}(t)$, is a (point-in-space valued) function that specifies the location of the particle at time t. Of course, there are many possible different motions along the curve and thus many possible parametrizations. In the x-y plane $\mathbf{p}(t) = (x(t), y(t))$, and in 3-space with coordinates x-y-z we have $\mathbf{p}(t) = (x(t), y(t), z(t))$; but for geometric purposes it is best to think of $\mathbf{p}(t)$ as a point in space without specific coordinates.

The ***derivative*** of $\mathbf{p}(t)$ with respect to t or ***velocity of the parametrization*** is

$$\mathbf{p}'(t) = (x'(t), y'(t)) \text{ or } \mathbf{p}'(t) = (x'(t), y'(t), z'(t))$$

or, free of coordinates,

$$\mathbf{p}'(t) = \lim_{h\to 0}(1/h)(\mathbf{p}(t+h) - \mathbf{p}(t)).$$

Now, the graph of a function $y = f(x)$ has a natural parametrization $\mathbf{p}(x) = (x, f(x))$, and the derivative of this parametrization is the velocity vector $(1, f'(x))$. Notice that the velocity vector of this parametrization of the graph is never zero. We can also consider the graph of any parametrization. For example,

$$\mathbf{p}(t) = (r \cos t, r \sin t)$$

parametrizes a circle in the plane, and in 3-space its graph

$$\mathbf{q}(t) = (t, r \cos t, r \sin t)$$

represents a helix.

Geometrically, if we again imagine a particle or bug moving along the curve at a constant speed, we can talk about the ***tangent direction*** at a point on a curve as the direction of the motion at that point, and this makes sense without any reference to units. The tangent direction at $\mathbf{p}$ is the direction of the tangent vector at $\mathbf{p}$. In the plane, direction can be designated by the slope, but in 3-space it is more convenient in linear algebra (and more directly relevant to the geometry of a curve) to use unit vectors. The ***unit tangent vector*** is defined to be

$$\mathbf{T}(t) = \frac{\mathbf{p}'(t)}{|\mathbf{p}'(t)|}.$$

In general a vector has direction and magnitude (or length); in a unit vector we factor out the length in order to leave only direction.

If the velocity $\mathbf{p}'(t)$ is non-zero and h is small, then the direction from $\mathbf{p}(t)$ to $\mathbf{p}(t+h)$ approximates the tangent direction at $\mathbf{p}(t)$. Even better, the direction from $\mathbf{p}(t+h)$ to $\mathbf{p}(t-h)$ is almost always orders of magnitude better as an approximation to the tangent direction at $\mathbf{p}(t)$ than the direction from $\mathbf{p}(t+h)$ to $\mathbf{p}(t)$. This is clear from a picture, as in Figure 2.4.

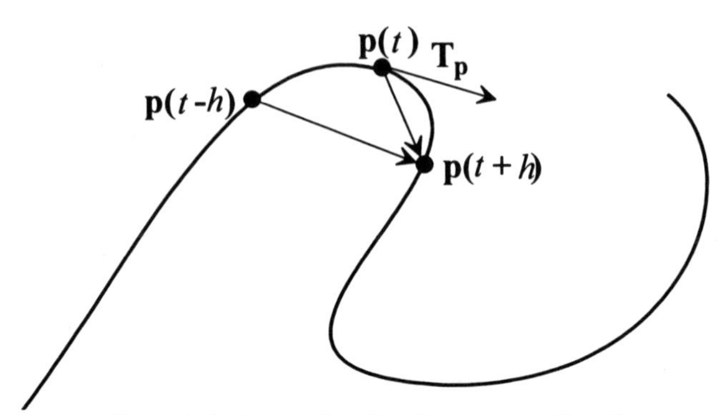

Figure 2.4. Approximating the tangent direction.

f. *Let* $\mathbf{x}(t)$ *be a parametrization of a curve in* $\mathbb{R}^n$*. What is the relationship between the derivative (velocity) existing and the curve being infinitesimally straight at* $\mathbf{p} = \mathbf{x}(t)$*? Under what conditions does the existence of a velocity vector imply that the curve is smooth? Why?*

Suggestions

Computer Exercise 2.2 allows you to use the computer to draw a curve with its tangent vectors displayed at specified points.

Start by looking at curves in the plane that are the graphs of functions. Then look at circles and other parametrized curves in the plane. Think of the parametrization as describing a motion along the curve. If the parametrization describes the motion of a small bug who is walking along the curve, then in what direction is the bug looking at any given time? If you walk along a path with a sharp corner in it, what happens to the parametrization of your motion when you get to the corner? Is it possible to have a smooth curve which has a parametrization that is not differentiable? Remember the difference between geometric curve, parametrized curve and the graph of a function.

The most natural geometric parametrization of a curve is a ***parametrization with constant speed***, which means that the magnitude of the velocity vector is a constant (and nonzero). When there is a unit of distance then we can have a ***parametrization with respect to arclength***, that is, we label points according to how far they are (along the curve) from some reference point. It is usual in differential geometry texts to use the letter s to denote a parameter with respect to arclength. For example,

$$\mathbf{p}(s) = (\, r\cos(s/r),\, r\sin(s/r)\,)$$

is a parametrization of the circle with respect to arclength. (**Be sure you see why!**) For the helix

$$\mathbf{q}(t) = (\, ht,\, r\cos t,\, r\sin t\,)$$

the velocity vector is

$$\mathbf{v}(t) = \dot{\mathbf{q}}(t) = (h,\ -r\sin t,\ r\cos t)$$

and the speed is

$$|\mathbf{v}(t)| = \frac{ds}{dt} = \sqrt{h^2 + r^2}$$

and thus

$$s = t\sqrt{h^2 + r^2}$$

and therefore

$$s \to \left(\frac{hs}{\sqrt{h^2 + r^2}},\ r\cos\frac{s}{\sqrt{h^2 + r^2}},\ r\sin\frac{s}{\sqrt{h^2 + r^2}} \right)$$

is the same helix parametrized by arclength.

In addition, if $\mathbf{p}$ is a point on a curve it is sometimes convenient to talk about $\mathbf{p}+h$ as the point on the curve, which is a distance h away from $\mathbf{p}$ as measured along the curve. (See Figure 2.8 on page 53.)

Notice that if a curve is parametrized with respect to arclength or is parametrized with constant (nonzero) speed, then the curve is infinitesimally straight whenever the velocity vector exists. Moreover, we can prove that

A constant speed curve is smooth if and only if the velocity vector exists and is continuous at each point.

Problem 2.3. Curvature of a Curve in Space

a. *What ways can you think of to quantify the curvature at a point on non-straight curves? Find a method of quantifying curvature that you can conveniently apply to find the curvature at a point on a physical curve.*

Think of the curvature at a point both in terms of how much the curve is turning at the point and also in which direction it is curving. As examples, use circles, helixes, and physical curves that you make out of wire. How is curvature affected by different fields of view?

Since the curvature at a point **p** has both magnitude and direction it is often represented as a vector and denoted by $\kappa_{\mathbf{p}}$ or $\kappa(\mathbf{p})$ or (when **p** is understood) simply as κ.

b. *Compare your method(s) with the following classical definition: The curvature is the rate of change at* **p** *of the unit tangent vector* **T** *with respect to arclength, that is,*

$$\kappa = \frac{d\mathbf{T}}{ds}, \ \textit{or} \ \ \kappa(p) = \lim_{h \to 0} \frac{\mathbf{T}(p+h) - \mathbf{T}(p-h)}{2h}.$$

c. *Why is the curvature vector always perpendicular to the tangent direction?*

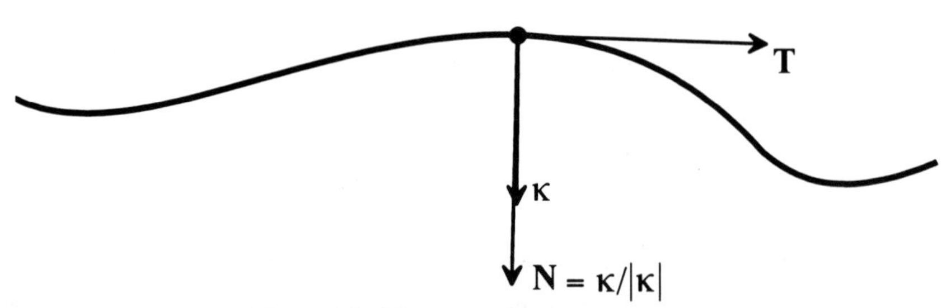

Figure 2.5. The normal vector to a curve.

In the plane there are two directions perpendicular to the tangent direction and, in higher dimensions, infinitely many such perpendicular directions. When the curvature vector is defined and nonzero, then we can denote by **N** (called the ***normal vector*** to the curve) the unit vector in the direction of the curvature vector, i.e. $\mathbf{N} = \kappa/|\kappa|$. (See Figure 2.5.)

d. *For a planar curve, the magnitude of the curvature is equal to the magnitude of the rate of change at* **p** *of the normal vector* **N**, *that is,*

$$|\kappa| = \left|\frac{d\mathbf{N}}{ds}\right|.$$

Why is "planar" important?

Suggestions

Computer Exercise 2.3 allows you to have the computer draw a curve with its curvature vectors displayed at certain points.

Again it may be helpful to think of a bug crawling along the curve. If its motion is parametrized by arclength, how is it crawling? How can you find the circle that passes through three given points on a curve? Play with a piece of wire. Also, use the parametrization given above of the helix (with respect to arclength),

$$\mathbf{p}(s) = \left(\frac{hs}{\sqrt{h^2 + r^2}}, r\cos\frac{s}{\sqrt{h^2 + r^2}}, r\sin\frac{s}{\sqrt{h^2 + r^2}}\right),$$

where h, r are constants, and calculate the curvature vector and the derivative of the normal vector.

Implicit in Problem **2.3** is the Lemma:

The derivative of a unit vector (a geometric direction) is always in a direction perpendicular to the original direction.

If the unit vector were to change in a direction that is not perpendicular to its own direction, then the length of the vector would change. It is important at this point to be able to see the truth of this statement geometrically. But it can also be useful to understand the linear algebra proof that goes like this:

If $\mathbf{V}(s)$ is a vector-valued function of the real variable s such that $|\mathbf{V}(s)| = \text{constant}$, then writing

$$|\mathbf{V}(s)|^2 = \mathbf{V}(s)\cdot\mathbf{V}(s)$$

and differentiating with respect to s and using the product rule, we obtain

$$\mathbf{V}'(s)\cdot\mathbf{V}(s) + \mathbf{V}(s)\cdot\mathbf{V}'(s) = 0$$

and thus $\mathbf{V}'(s)$ is perpendicular to $\mathbf{V}(s)$, since the dot product commutes.

Note that the curvature need not vary continuously even when the curve is infinitesimally straight everywhere. For example, join two circular arcs of different radii or join straight segments to a circular arc such as in Figure 2.6.

If the curve on the right is the cross-section of the edge of a table top, then the discontinuities in the curvature can actually be felt with one's finger. If it is the cross-section of a boat, then the discontinuities in the curvature cause water turbulence and friction. It is also difficult to drive at high speeds on a road with discontinuities in the curvature, because at the discontinuity the steering wheel must be instantaneously turned.

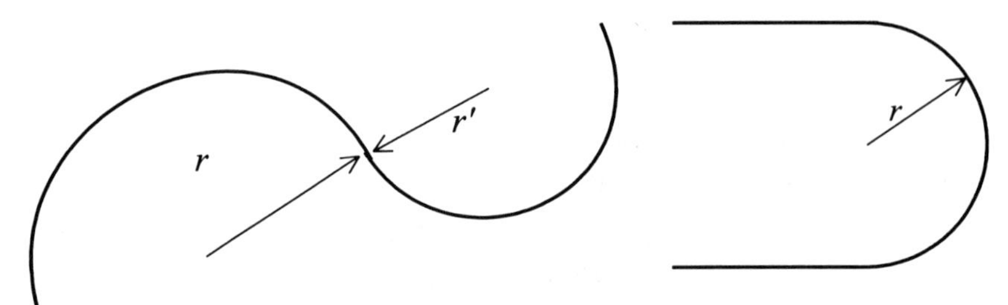

Figure 2.6. Curvature need not vary continuously.

When we say that a curve has ***constant curvature*** we mean the scalar curvature $\kappa = |\boldsymbol{\kappa}|$. This is the only meaning we could have because:

If the curvature is not zero then the curvature vector is never constant (because its direction must be changing).

[Be sure you see why this is true.]

Curvature of the Graph of a Function

THEOREM. *If $(x, f(x))$ is the graph of a twice differentiable function, then at the point $(a, f(a))$ we have*

$$\mathbf{T}(a) = \frac{(1, f'(a))}{\sqrt{1 + (f'(a))^2}}$$

and

$$|\kappa(a)| = \frac{|f''(a)|}{[1 + (f'(a))^2]^{3/2}}.$$

Proof. The graph is a curve $\gamma(x) = (x, f(x))$. Note that the velocity vector

$$\gamma\,'(x) = (1, f'(a))$$

is never zero and thus, $\mathbf{T}(a)$ is the unit vector in the direction of the velocity vector:

$$\mathbf{T}(a) = \frac{\gamma'(a)}{|\gamma'(a)|} = \frac{(1, f'(a))}{\sqrt{1 + (f'(a))^2}}.$$

Now the curvature is the rate of change of $\mathbf{T}$ with respect to arclength

$$s = \int_0^a |\gamma'(x)|dx = \int_0^a \sqrt{1 + (f'(x))^2}\, dx,$$

thus,

$$\frac{ds}{dx} = \sqrt{1 + (f'(x))^2}\,.$$

A geometric interpretation of this last formula is shown in Figure 2.7. From the figure we see that, in a f.o.v. where the graph is indistinguishable from the tangent line,

$$(ds)^2 = (dx)^2 + (dy)^2 = (dx)^2 + (f'dx)^2 = \left(1 + (f')^2\right)(dx)^2,$$

from which the desired formula follows.

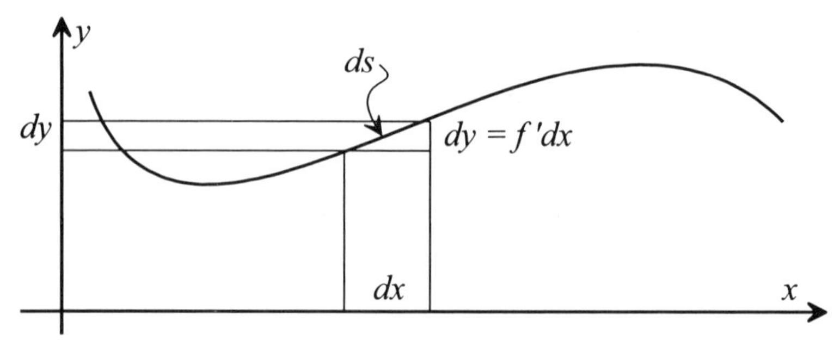

Figure 2.7. Element of arclength.

We can now calculate the curvature vector:

$$\kappa(a) = \frac{d\mathbf{T}}{ds} = \frac{d\mathbf{T}}{dx}\frac{dx}{ds} = \frac{d}{dx}\left[\frac{(1, f'(x))}{\sqrt{1+(f'(x))^2}}\right]_{x=a}\left(\frac{ds}{dx}\right)^{-1}_{x=a} =$$

$$= \left[\frac{f''(a)}{[1+(f'(a))^2]^{3/2}}(-f'(a), 1)\right]\left(\frac{1}{\sqrt{1+(f'(a))^2}}\right)$$

and thus the theorem follows. Note that at a point at which

$$f'(a) = 0, \quad \kappa(a) = f''(a)\,(0,1).$$

Problem 2.4. Osculating Circle

If γ is a curve with nonzero curvature $\kappa_{\mathbf{p}}$ at the point **p**, then the curve's ***osculating circle at*** **p** is the circle $C_{\mathbf{p}}$ through **p** which has the same curvature vector and unit tangent vector as γ. (See Figure 2.8.)

a. *Show that the osculating circle lies in the plane determined by the curvature* $\kappa_{\mathbf{p}}$ *and unit tangent vector* $\mathbf{T}_{\mathbf{p}}$ *and that its radius is* $r = 1/|\kappa_{\mathbf{p}}|$.

The plane of the osculating circle is called the ***osculating plane at*** **p**. You may find it helpful at this point to try Computer Exercise 2.4, which allows you to display a curve with its osculating planes (or osculating circles) displayed at various points. The radius, r, of the osculating circle is called the ***radius of curvature at*** **p** and the curvature vector has magnitude (length) $1/r$.

As we zoom in (with fixed tolerance) on the point **p**, the circle and the curve will both become indistinguishable from the tangent line—this is what we mean by being tangent. If the curve is locally straight at **p**, then in all f.o.v.'s, the curve and tangent line coincide and there is no osculating circle. But if the curve is not locally straight at **p**, then by decreasing the tolerance, if necessary, we will be able to distinguish the curve from the tangent line. In that f.o.v., there is an approximation of the osculating circle. We may pick two (not collinear with **p** in the f.o.v.) points, **p**–h and **p**+h, on either side of **p** at a distance h from **p** along the curve. Then the circle determined by **p**–h, **p**, **p**+h approximates $C_{\mathbf{p}}$. We see in Figure 2.8 that the line through **p**–h and **p** approximates the tangent line at **p**–h/2, and the line through **p** and **p**+h approximates the tangent line at **p**+h/2. Let $\Delta\mathbf{T}$ be the change in the unit tangent vector from **p**–h/2 to **p**+h/2.

b. *The rate of change (with respect to arclength) of the unit tangent vector is approximated by* $|\Delta\mathbf{T}/h|$, *which is in turn approximated by* α/h. *In the limit*

$$|\kappa| \equiv \left|\frac{d\mathbf{T}}{ds}\right| \equiv \lim_{h\to 0}\left|\frac{\Delta\mathbf{T}}{h}\right| = \lim_{h\to 0}\frac{\alpha}{h} = \frac{1}{r},$$

and then ***the osculating circle at*** **p** *is the limiting circle of radius r.*

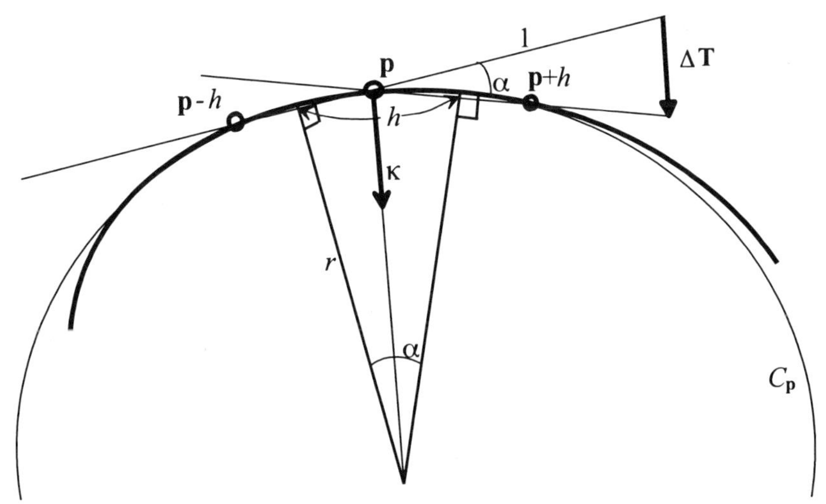

Figure 2.8. Approximating the osculating circle.

Notice that the above discussion gives us another way of defining the curvature of a curve in a plane.

c. *If* γ *is a planar curve parametrized by arclength and* $\theta(\mathbf{p})$ *is the angle between* $\mathbf{T}_\mathbf{p}$ *and some fixed direction, for every point* $\mathbf{p} = \gamma(s)$, *then*

$$|\kappa| = \left|\frac{d\theta}{ds}\right|.$$

Notice that, since the angle between vectors does not depend on their length, θ *can be considered the angle between the tangent direction at* **p** *and the fixed direction.*

If the curvature is zero at a point, then the tangent line approximates the curve better than any circle and there is no osculating circle or osculating plane. If the curve has zero curvature at every point along an interval of the curve, then that interval is straight (there is no turning), but if the curvature is merely zero at a point, then the curve is not straight at that point (examples include inflection points and the vertex of the graph of $y = x^4$.)

***d.** *Consider a segment of a curve that is indistinguishable from the osculating circle. If we move this segment along the normal vector at each point, then its arclength* l *will change at the rate* $-|\kappa|\, l$.

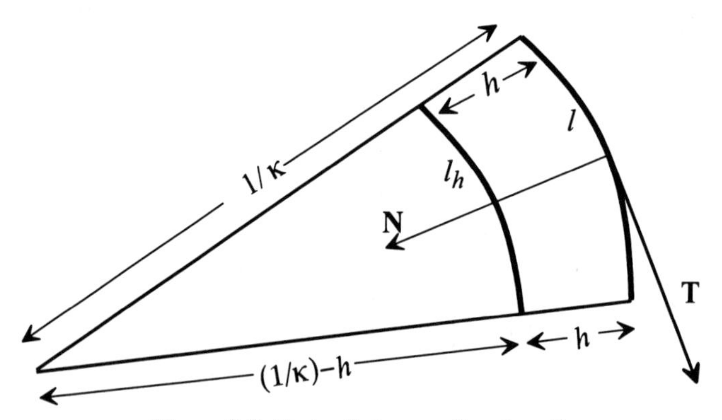

Figure 2.9. Rate of change of arclength.

[Hint: In Figure 2.9 the reader can show that the derivative

$$\frac{d}{dh}l_h = -|\kappa|l.]$$

PROBLEM 2.5. Strakes

To help us understand the idea of curvature of curves and, later, the curvature of surface, let us look at the following example.† To give structural support to large metal cylinders, such as large smoke stacks, engineers sometimes attach a spiraling strip called a ***strake***. See Figure 2.10.

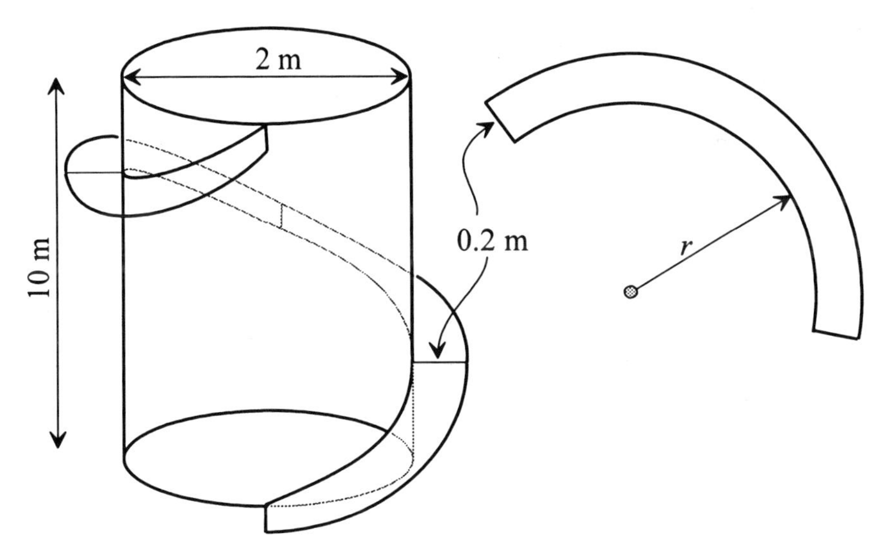

Figure 2.10. A strake.

To produce the strake it is convenient to cut annular pieces from a flat sheet of steel as illustrated in Figure 2.10. These annular pieces are then bent along the helix to form the strake.

a. *Convince yourself that the way to compute the ideal value for the radius r is to require that the helix on the cylinder and the inner curve of the annulus have the same curvature.*

[Hint: What does "curvature" mean?]

†This example is inspired by an example in [**DG**: Morgan, pp.6-10].

b. *Compute this ideal value for r.*

[Hint: Use the formulas given in Problem **2.3**.]

c. *Can the flat annulus exactly fit a piece of the strake?*

[Hint: Clearly the strake is not planar as it stands. But can the annular piece of steel be bent without stretching in order to produce the strake? We are asking if the strake is ***locally isometric*** to the plane, or, in other words, if the local *intrinsic* geometry of the strake is the same as the local geometry on a plane. Cut a paper annulus with the ideal radius and try forming it into a strake. Consider the inner and outer edges of the strake and the annulus. (Notice that both the outer and inner edges of the strake are helixes.) You might compute the curvatures and lengths of the inner and outer edges of the annulus and the corresponding inner and outer edges of the helical strake. What do you conclude? What happens if we make the strake very wide compared to the diameter of the cylinder—such as happens in an auger (see Figure 2.11)?]

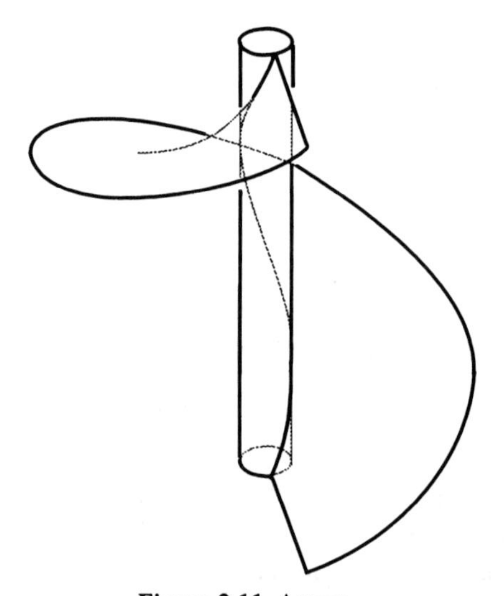

Figure 2.11. Auger.

PROBLEM 2.6. When a Curve Does Not Lie in a Plane

Suppose we have a smooth curve that does not lie in a plane and that has a well-defined (nonzero) curvature vector at each point. Then at some point on the curve the osculating plane (the plane of the osculating circle) must be changing. Let us make a picture of how the osculating plane could be changing at such a point **p**. If the osculating plane is changing then there must be points, **p**– and **p**+, close on either side of **p** with different osculating planes. Let us use the three point technique to draw approximations of these osculating planes [**q**, **p**+, **p**– determine the plane containing **a**, and **p**+, **p**–, **r** determine the plane with **b**] as in Figure 2.12.

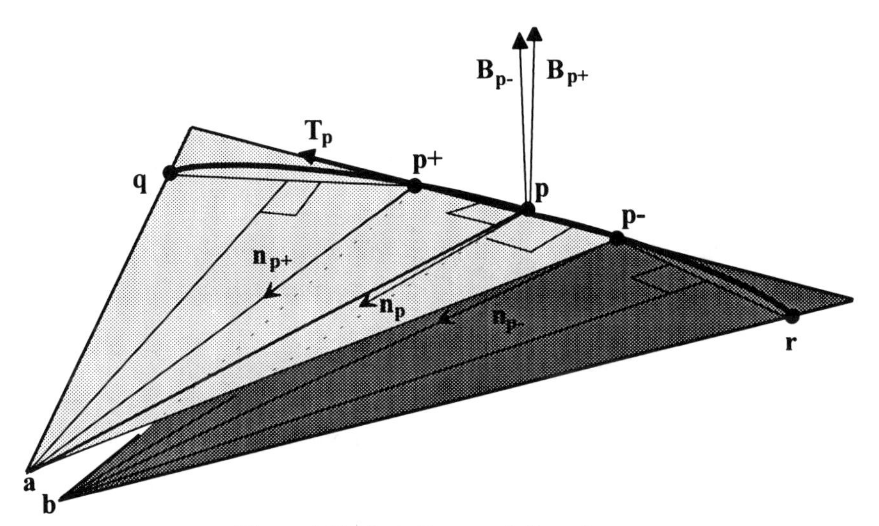

Figure 2.12. Rotating osculating planes.

Notice that the two approximate osculating planes intersect on the cord from **p**– to **p**+ and that the three points, **p**, **a**, **b**, lie in a plane that is perpendicular to this cord, which closely approximates $\mathbf{T}_p$.

a. *In the situation described in the above two paragraphs, argue that at* **p** *the osculating planes must be pivoting around the tangent line* $\mathbf{T}_p$ *and that the centers of curvature are changing. Also show that they are changing in the plane perpendicular to* $\mathbf{T}_p$.

[Hint: Study Figure 2.12. If you are having trouble seeing what is happening, then make a model using two sheets of paper and tape.]

A unit vector perpendicular to the osculating plane at **p** is called the ***binormal***, $\mathbf{B}_p$. Note that the unit tangent vector, normal vector, and binormal vector $\{\mathbf{T}_p, \mathbf{N}_p, \mathbf{B}_p\}$ are all unit vectors, which are mutually orthogonal and thus form an orthonormal basis (see Appendix A.3) for $\mathbb{R}^3$. We pick the direction of $\mathbf{B}_p$ by specifying that $\{\mathbf{T}_p, \mathbf{N}_p, \mathbf{B}_p\}$ be ***right handed***, that is, if you curl the fingers of your right hand from $\mathbf{T}_p$ to $\mathbf{N}_p$, then your thumb will point in the direction of $\mathbf{B}_p$. In terms of the cross product, $\mathbf{B}_p = \mathbf{T}_p \times \mathbf{N}_p$. The three unit vectors, $\mathbf{T}_p$, $\mathbf{N}_p$, $\mathbf{B}_p$, are called the ***Frenét frame*** at the point **p** and, of course, they vary from point to point; but, at each point, they form an orthonormal basis for $\mathbb{R}^3$. You may find it useful at this point to use Computer Exercise 2.6 to display a curve with its Frenét frames displayed at various points.

Most books define the ***torsion*** (vector) of a curve to be the rate of change (with respect to arclength) of the binormal, in symbols $\tau_p = \mathbf{B}_p'$. So we can:

b. *Conclude that a curve with well-defined (nonzero) curvature is planar if and only if the binormal is constant (or, its torsion is everywhere zero).*

Note that "nonzero" is necessary since a smooth plane curve that is composed of two arcs of circles put together like a figure S (see, for example, Figure 2.6 above) has different binormals on the two different arcs.

You may use part **a** to show part **b** or you may find it helpful to use some version of the Mean Value Theorem from first semester calculus. However, be warned that the Mean Value Theorem applies, in general, only to differentiable real-valued functions of one real variable or to differentiable curves in the plane.

For example, one way of stating a ***Mean Value Theorem for Planar Curves*** is:

Given two points **p** *and* **q** *on a differentiable curve in the plane, there is some point* **r** *on the curve between* **p** *and* **q** *such that the tangent vector* $\mathbf{T}_r$ *at* **r** *is parallel to* $\mathbf{p} - \mathbf{q}$. (See Figure 2.13.)

This theorem applies to real-valued functions of one real variable if you consider the graph of the function as a curve in the plane. Note that this result is not true, in general, for curves that do not lie in a plane; for example, the reader can easily find two points on a helix for which it does not hold. (See Problem **4.2**.)

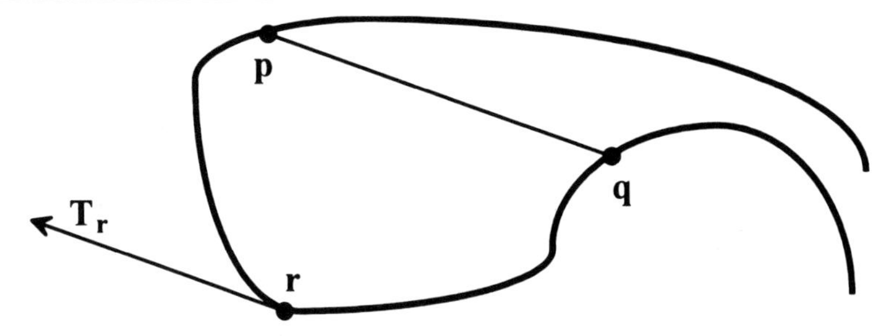

Figure 2.13. Mean Value Theorem for Plane Curves.

When you compress a helical spring so that it becomes more nearly planar, the curvature stays constant (and thus the diameter increases), but the torsion decreases.

This discussion leads to equations that express the connections between the rate of change (with respect to arclength) of the three unit vectors, **T**, **N**, **B**. These equations, called the Frenét-Serret Equations, were independently found by Fréderic-Jean Frenét (1847, published in 1852) and Joseph Serret (1851):

***c.** *Prove the following **Frenét-Serret Equations**:*

$$\begin{aligned} \kappa \equiv \mathbf{T}'(s) &= \kappa(s)\mathbf{N}(s), \\ \mathbf{N}'(s) &= -\kappa(s)\mathbf{T}(s) + \tau(s)\mathbf{B}(s), \\ \tau \equiv \mathbf{B}'(s) &= -\tau(s)\mathbf{N}(s), \end{aligned}$$

where $\kappa(s) = |\mathbf{T}'(s)|$ *is the **scalar curvature** and*

$$\tau(s) = -[B'(s) \cdot N(s)]$$

*is the **scalar torsion**.*

The definition of

$$\tau(s) = -\mathbf{B}'(s)\mathbf{N}(s)$$

is a sign convention (the most commonly used one), but other conventions are possible and are used in some books.

You may find it easiest initially to prove the first and third equations by using Problem **2.3** and the drawing above. These two equations can then be combined to obtain the second equation by either looking geometrically at the three mutually perpendicular unit vectors or by differentiating the cross product **N** = **B** x **T**. Note that from the right-hand rule we can conclude that, if **B** = **T** x **N**, then **T** = **N** x **B** and **N** = **B** x **T**.

We will not use the Frenét-Serret Equations in this text, but they are a tool commonly used to study curves in 3-space. See, for example, [**DG**: Millman/Parker, Chapter 2]. In that book is proved [Theorem 5.2]:

> *Any smooth curve with nonzero curvature is completely determined, up to position, by its curvature and torsion.*

d. *Calculate the binormal and torsion of a helix. Show that its scalar torsion is constant.*

In higher dimensions the above discussion of **T**, κ, and **N** hold without change. In addition, three (non-collinear) points determine a circle and a plane in any dimensions, and thus the notion of osculating circle and osculating plane makes sense in all dimensions. However, in higher dimensions, planes are not determined by a normal vector, and thus the binormal vector does not make sense and, when needed, must be replaced by tensors or forms.

Chapter 3

Extrinsic Descriptions of Intrinsic Curvature

When we view a curve on a smooth surface, then we can talk about its ***intrinsic curvature*** (sometimes called ***geodesic curvature***) with respect to the surface. As in Chapter 2, we want the curvature to be the rate of change of the tangent direction (with respect to arclength), except now we want to look at those changes that are intrinsic to the surface (that is, the changes that a 2-dimensional bug on the surface would be able to detect by perceiving only what is on the surface). The notions from Problem **2.3** can be used, but we have to be careful. For example, the machinery of ordinary linear algebra does not directly allow us to compare tangent vectors at two different points on the surface. This is because the vectors tangent to the surface do not, in general, form a vector space, since the difference of two vectors tangent to a surface at different points is not necessarily also a vector tangent to the surface. *Check this out on small portions of a cylinder and sphere.* (See discussion before Problem **2.2**.) We don't know how to talk about vectors at two different points on a sphere as being the "same". North/South/East/West terminology does not work on the sphere because these directions depend on the choice of pole and because at the north pole every direction is south.

On a surface, a curve with no intrinsic curvature is said to be ***intrinsically straight*** (with respect to that surface) or is called a ***geodesic***. These are the curves that a 2-dimensional bug would experience as straight. Along a geodesic, the tangent direction is not changing intrinsically, but of course, in general, it is changing extrinsically. So, we want to be able to find a way of talking about two tangent directions being equal intrinsically (along a curve) when they are not necessarily equal extrinsically—this we will do in Chapter 5.

In this chapter we give extrinsic descriptions of intrinsic curvature. These descriptions make sense to us viewing the surface extrinsically but are of no use to the 2-dimensional bug. Nor will they be useful to us in our experience as intrinsic "bugs" in our 3-dimensional physical universe where we have no extrinsic experience. We will remedy this situation later in Chapter 5 where we will finally arrive at intrinsic descriptions.

Problem 3.1. Smooth Surfaces and Tangent Planes

An (***extrinsic***) ***smooth surface***, M, is a geometric figure in $\mathbb{R}^n$ that is uniformly infinitesimally planar (or flat). We say that a surface is *infinitesimally planar* at the point p in M if, when you zoom in on p, then M will become indistinguishable from a plane, T_pM, called the ***tangent plane at*** p; that is, for every tolerance $\tau > 0$, there is a δ, such that in any f.o.v. centered at p of radius $< \delta$, it is the case that the projection of M onto T_pM is one-to-one and moves points less than $\tau\delta$ in the f.o.v. [Remember that the tolerance is a percentage such that two points are indistinguishable (in the f.o.v.) if their distance apart is less than the tolerance times the diameter of the f.o.v. (See Problem **2.1**.)] The surface is uniformly infinitesimally planar if there is some neighborhood of p such that (for each tolerance τ) the same δ can be used for each point in the neighborhood. Computer Exercise 3.1 may be used to view a surface given parametrically and its tangent plane at a specified point. (See Problem **4.1** for further discussion of the tangent plane.)

***a**.*[†] *Show that a surface is smooth if and only if it is infinitesimally planar and the tangent planes, T_pM, vary continuously with respect to p.* (That is, for every tolerance τ there is a field of view with center p and radius ρ, such that, if p and x are both on the surface in the field of view, then in the field of view each point on T_p is within $\tau\rho$ of the tangent plane T_x.)

[Hint: Use the ideas from Problem **2.2.e**.]

b. *Let M be a surface with a coordinate patch* $\mathbf{x}(x,y)$. *If there exist continuous partial derivatives of* $\mathbf{x}(x,y)$ *that are linearly independent, then M is a smooth surface. Give an example to show that the converse is false.*

[Hint: Use Part **a**.]

[†]Problems and Sections marked with an asterisk (*) are not essential later in this book.

c. *Show that cylinders, spheres, and strakes are smooth surfaces and that a cone is a smooth surface except at the cone point.*

d. *If the function f is nonzero, then show that a surface of revolution*

$$\mathbf{x}(\theta, x) = (x, f(x) \cos \theta, f(x) \sin \theta)$$

is a smooth surface if and only if the function f is smooth.

[Hint: Use Part **b** and Problem **2.2**.]

***e.** *Show that the graph of $g(x, y)$ is a smooth surface with every tangent plane projecting one-to-one onto the (x, y)-plane if and only if the function g has partial derivatives which exist everywhere and are continuous* (that is, g is C^1).

[Hint: Use the ideas from Problem **2.2**.]

We can now prove the following extension of Problem **1.9**:

COROLLARY. *If a smooth surface is the graph of a function, then the function is continuously differentiable.*

f. *Show that each point on the annular hyperbolic plane has a neighborhood that can be isometrically embedded into 3-space as a smooth surface.*

[Hint: See Problem **1.8** for the description of the annular hyperbolic plane made from annular strips. First, argue that each point on the annular hyperbolic plane is like any other point. Second, start with one of the annular strips and complete it to a full annulus in a plane. Third, construct a surface of revolution by attaching to the inside edge of this annulus other annular strips as described in the construction of the annular hyperbolic plane. Fourth, imagine the width of the annular strips, δ, shrinking to zero. (See Figure 3.1.)]

It is a theorem (see Section 5-11 in [**DG**: do Carmo]) that there is no way to embed isometrically the whole complete hyperbolic space as a *smooth* surface in 3-space. Unfortunately, many books quote this result without the word "smooth" because it has been traditional to assume that everything is smooth. Thus it follows that no matter how you isometrically position the annular hyperbolic plane in 3-space, there must be points that have no tangent plane. If you play with a paper model you can see this phenomenon.

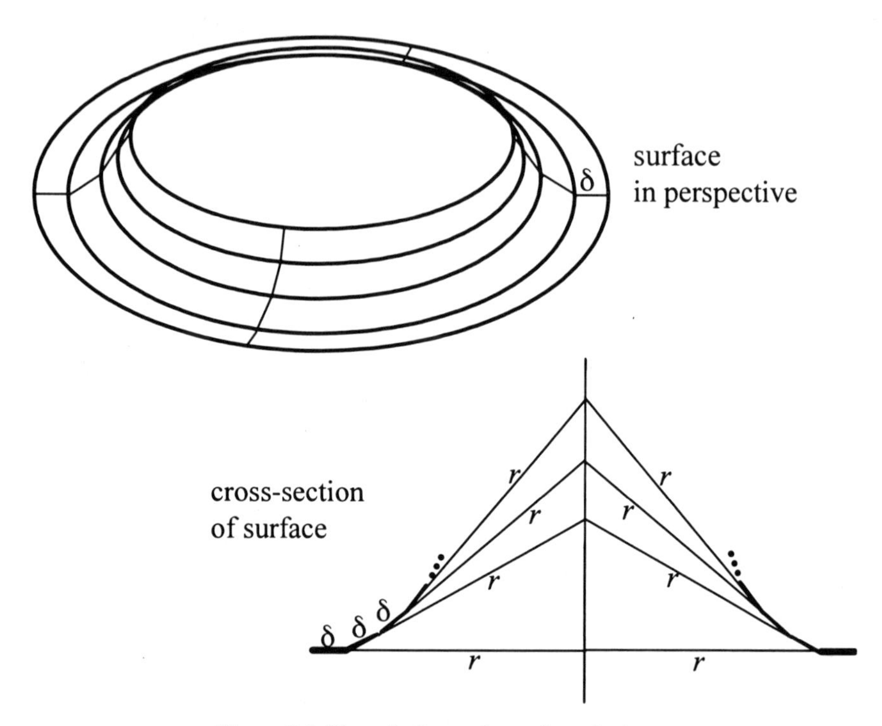

Figure 3.1. Hyperbolic surface of revolution.

The ***normal space*** $N_p M$ at a point p in the smooth surface M is the union of all lines which are perpendicular to $T_p M$ at p. In 3-space the normal space is a line called the ***normal line***.

PROBLEM 3.2. Extrinsic Curvature: Geodesics on Sphere

We now will start to explore curves on smooth surfaces and investigate the relationship between their (extrinsic) curvature vectors and the normal vectors and tangent planes of the surface. You may find two computer exercises helpful. Computer Exercise 3.2a allows you to display a surface and a curve in that surface. Computer Exercise 3.2b allows you to display extrinsic curvature vectors of this curve.

a. *Show that in every case the extrinsic curvature of any geodesic on a cylinder or cone is pointing in a direction that is normal to the surface.*

b. *Convince yourself that for any geodesic on the sphere the extrinsic curvature (or normal vector) must point towards (or directly away from) the center of the sphere.*

[Hint: If the extrinsic curvature were not so aligned, what would be the intrinsic experience of moving along the curve?]

c. *Prove that no other curve is a geodesic on the sphere except for an arc of a great circle.*

[Hint: You may find it helpful to use Problem **2.6.a-b** (*What Happens When a Curve Does Not Lie in a Plane?*), since great circles are precisely those curves on the sphere that are the intersection of the sphere with a plane that passes through the center of the sphere.]

PROBLEM 3.3. Intrinsic Curvature: Curves on Sphere

a. *How much of what you said about curvature in Problem* **2.3** *will hold for curves on a cylinder or on a cone?*

[Hint: Remember that some of what we did in Problem **2.3** used Euclidean geometry; and the theorems of Euclidean geometry do not, in general, hold on nonplanar surfaces. However, note that locally the cylinder and cone (away from the cone point) are intrinsically the same as (locally isometric to) a local portion of a Euclidean plane.]

b. *What is the intrinsic curvature of the latitude circle that forms an angle of* α *with the equator* (see Figure 3.2)*? Which latitudes have no intrinsic curvature?*

[Hint: Remember that a sphere is not locally isometric to the plane, but we can look at the cone that is tangent to the sphere along the latitude. Argue that a 2-dimensional bug will experience the intrinsic curvature of the latitude on the sphere as the same curvature as the intrinsic curvature of the latitude on the cone.]

Note that the latitude circle in Figure 3.2 has four different centers: The ***extrinsic center*** (or ***center of extrinsic curvature***) is the point **c** in the plane of the latitude at the center of the circle. The ***intrinsic center*** is the point **b**, which is the center of the circle with respect to the surface of the sphere—it is the center of the circle from the point of view of a 2-dimensional bug on the surface. Then there is the point **a**, which is the intersection of all the planes tangent to the sphere along the latitude circle. The point **a** can be called the ***center of intrinsic curvature***. Do you see why this name makes sense? What happens to the curve of tangency when you open up the cone? The point **d** is the center of the sphere and is the ***center of normal curvature*** for the latitude circle.

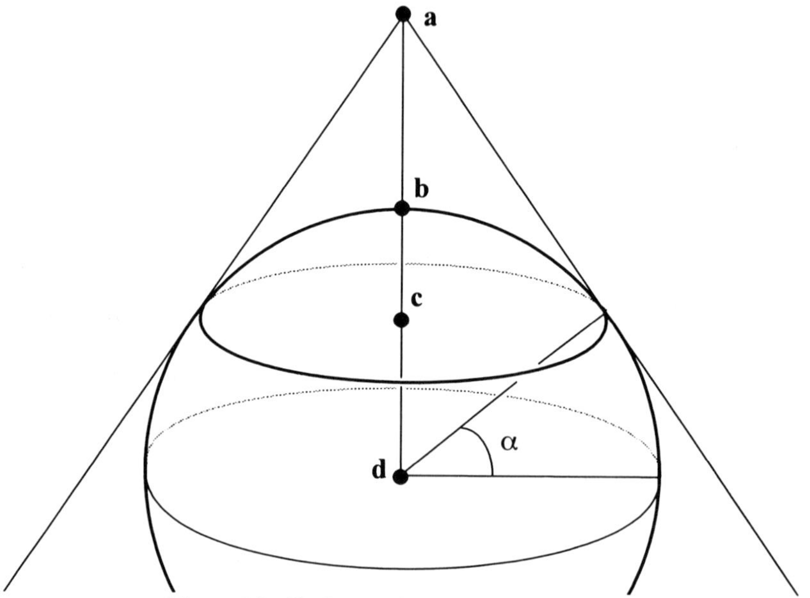

Figure 3.2. Circle on sphere with four centers.

c. *Show that, for all curves on cylinders, cones, and spheres that are extrinsic circles, the intrinsic curvature is the (orthogonal) projection onto the tangent plane of the extrinsic curvature vector.*

d. *Consider the extrinsic curvature vector at point p of a latitude circle on a sphere of radius R. Show that the projection of this vector onto the normal line to the sphere at p has length equal to 1/R. How do you make sense out of the fact that all these projections have the same length?* (See Figure 3.3.)

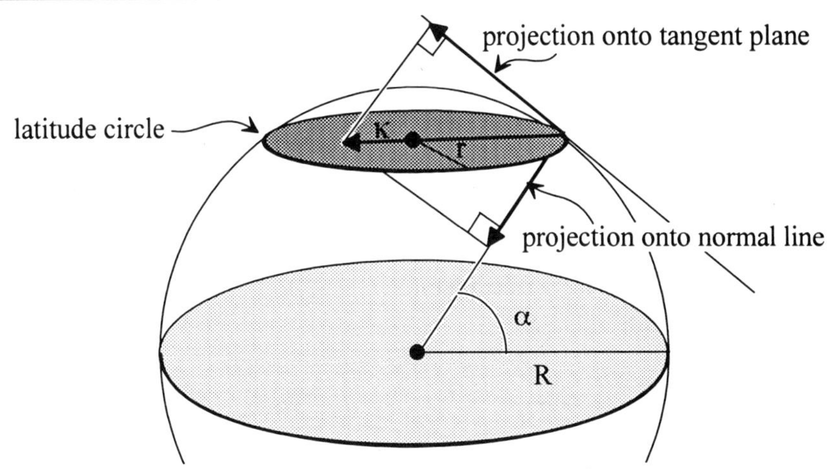

Figure 3.3. The three curvatures on a sphere.

Intrinsic (Geodesic) Curvature

In Problems **3.2-3.3** we saw that a curve γ on a surface S could be called a geodesic on S if and only if the curvature of γ is entirely due to the curvature of the surface S and not due to the intrinsic curving of the curve within the surface. We are, in effect, defining intrinsic curvature as the curvature observed by the bug. On the cone and cylinder we can define this precisely by flattening the cone or cylinder and then finding the curvature in the plane. We now wish to find a more formal definition of intrinsic curvature that will work on all surfaces. For now all that we can do is to find a formal *extrinsic* definition of the intrinsic curvature by defining the intrinsic curvature to be the projection of the extrinsic curvature onto the tangent plane. We eventually want to get intrinsic definitions because when we look at our own experiences in our three-dimensional universe (ignoring time for now) we have no access to extrinsic descriptions. But these intrinsic definitions will have to wait until later chapters.

We want to subtract out the component of the curvature of the curve that is perpendicular (normal) to the surface because this curvature is the curvature due to the surface. Thus we need to express this normal curvature.

This discussion leads to the following definitions. If γ is a smooth curve on the smooth surface S in $\mathbb{R}^n$, then let $\mathbf{T}$ and $\boldsymbol{\kappa}$ be the unit tangent vector and curvature vector to the curve at a point p on the curve. Define the ***intrinsic curvature*** *(or* ***geodesic curvature****)* of γ in S to be

$$\kappa_g = [\text{projection of } \kappa \text{ onto the tangent plane } T_pS\,]$$

and define the ***normal curvature*** of γ in S to be

$$\kappa_{\mathbf{n}} = [\text{projection of } \kappa \text{ onto the normal space } N_pS\,].$$

(See Figure 3.4.) This same picture except for a latitude circle on a sphere is given in Figure 3.5.

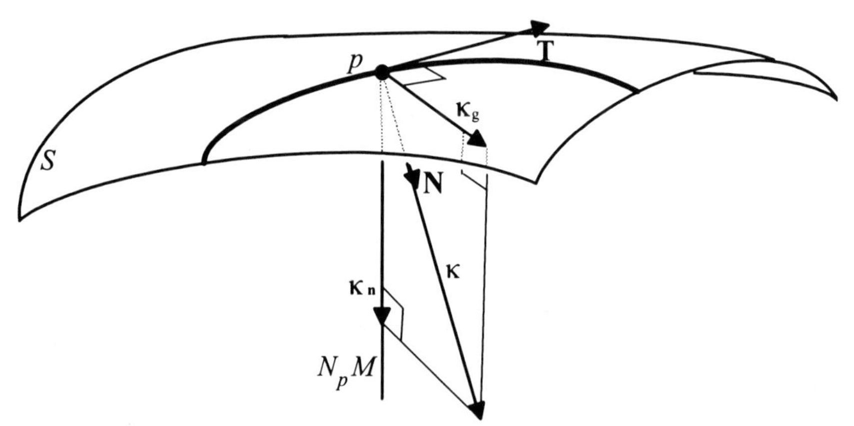

Figure 3.4. The three curvatures of a curve in a smooth surface.

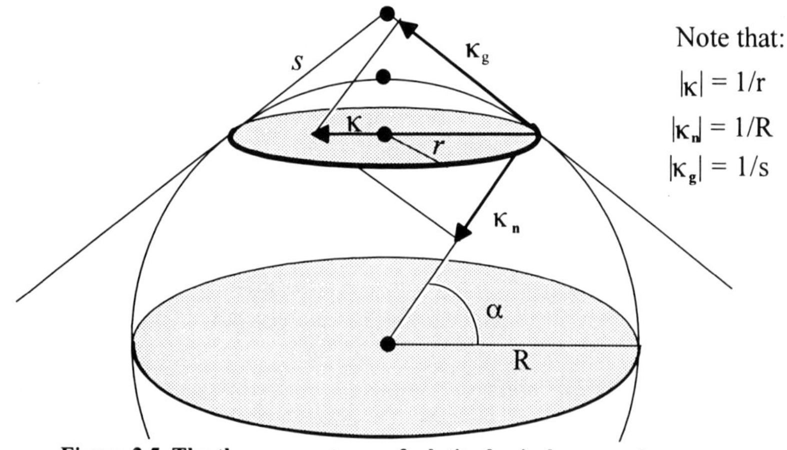

Figure 3.5. The three curvatures of a latitude circle on a sphere.

Computer Exercise 3.3 allows you to display a curve on a surface and the extrinsic, normal, and intrinsic curvature vectors of the curve at a specified point.

The directions of the three different curvatures, κ, κ_n, κ_g, give rise to three different normals: the extrinsic normal to the curve, the normal to the surface, and the intrinsic normal to the curve in the surface.

The curvature κ is the extrinsic curvature of the curve as a curve in space without reference to any surface containing it. However, the normal and intrinsic curvatures depend on the surface that one is considering. If the curve lies on two different surfaces then, in general, the normal and intrinsic curvatures will be different. Note that the extrinsic curvature is the vector sum of the intrinsic curvature and the normal curvature:

$$\kappa = \kappa_n + \kappa_g .$$

The curve γ is called a ***geodesic*** if and only if $\kappa_g = 0$ at *every* point. Note that a curve is a geodesic if and only if its extrinsic normal lies in the normal space N_pS at every point p. If $\kappa_g = 0$ at an isolated point, then we do not say it is a geodesic at that point. It is a geodesic at a point only if it has zero intrinsic curvature on a whole interval containing the point.

Problem 3.4. Geodesics on Surfaces — the Ribbon Test

a. *Show that if a curve γ on a surface is extrinsically a straight line, then it is a geodesic on the surface. Find as many examples of this as you can.*

b. ***Ribbon Test:*** *Consider a smooth embedding of a ribbon in* $\mathbb{R}^n$ (that is, the ribbon is a smooth surface) *such that the embedding is an isometry* (that is, distances measured along the ribbon do not change—no stretching).

i. *Argue intrinsically that the center line of the ribbon is a geodesic. What does this say about the extrinsic curvature vector κ?*

ii. *Use* **i.** *to show extrinsically that a thin ribbon "laid flat" on a smooth surface will always follow a geodesic. Here "laid flat" means that the ribbon is tangent to the surface along its center line.*

iii. *Use this ribbon test to find some geodesics on some physical smooth surfaces around your room.*

c. *Show* (using the definition above) *that, on a surface of revolution*

$$(r(z) \cos \theta, r(z) \sin \theta, z),$$

the curves with constant θ *are geodesics on the surface. Which generating circles* (z = constant) *are geodesics?*

d. *What geodesics can you find on the annular hyperbolic plane?*

[Hint: Use Parts **a** and **b** above and intrinsic symmetry.]

Note that a ribbon cannot be physically laid flat through a saddle point (where a curve on the surface that is perpendicular to the ribbon curves in the direction of the outward pointing normal). But if the ribbon is narrow with respect to the curvature of the surface, then it can be laid flat on the surface to a good approximation. To be precise with the ribbon test, we must imagine a ribbon that can pass through the surface so that it can be precisely tangent to the surface at every point along the centerline.

Ruled Surfaces and the Converse of the Ribbon Test

It is natural now to investigate whether the converse of the Ribbon Test holds: That is, if γ is a geodesic on a surface M, can a ribbon always be "laid flat" along γ? To answer this question we must first study a special·type of surfaces: ***ruled surfaces***.

Take a paper ribbon and bend it into many different smooth surfaces (no creases) in such a way that the center line is not extrinsically straight. Notice that at every point along the center line of the ribbon there is a direction along which the ribbon is (extrinsically) straight. It is important to notice that these directions of zero (extrinsic) curvature are not necessarily perpendicular to the center line and that these directions vary as you move along the center line. For example, lay the ribbon on a cylinder and on a cone and see which are the directions with zero extrinsic curvature (Figure 3.6).

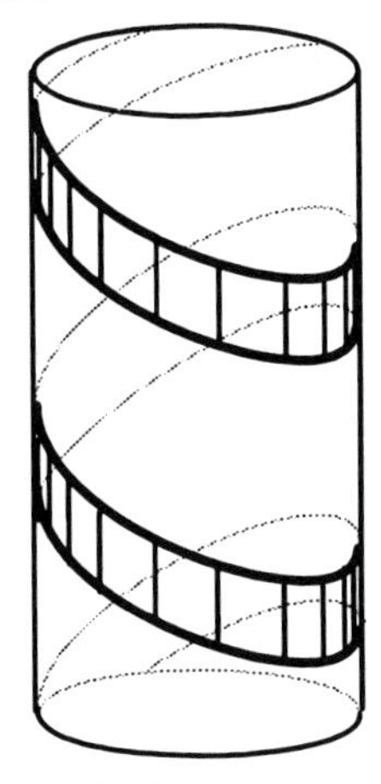
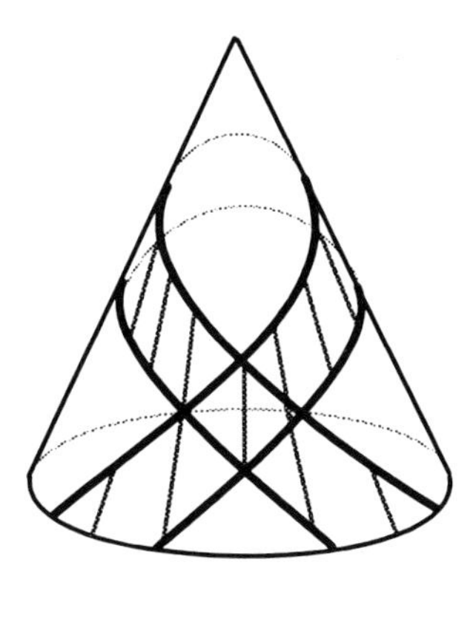

Figure 3.6. Rulings (segments with zero extrinsic curvature) along a ribbon.

A smooth surface M is called a ***regular ruled surface*** if on M there is a smooth curve $t \to \alpha(t)$ (parametrized by arclength) and at each point of the curve a unit vector $\mathbf{r}(t)$ such that

1. $\mathbf{r}(t)$ is a differentiable function of t,
2. each point $\alpha(t)$ is in the interior of an (extrinsically) straight segment in M that is parallel to $\mathbf{r}(t)$,
3. there is a (global) coordinate patch for M which can be expressed in the form: $\mathbf{x}(t,s) = \alpha(t) + s\mathbf{r}(t)$, and
4. the vectors, $\mathbf{x}_1(t,s) = \alpha'(t) + s\mathbf{r}'(t)$, $\mathbf{x}_2(t,s) = \mathbf{r}(t)$ form a basis for the tangent space.

The curve α is called the ***directrix*** of the surface, and the extrinsically straight segments are called the ***rulings*** of the surface.

Computer Exercise 3.4 allows you to display ruled surfaces.

Examples of ruled surfaces include cones (away from the cone point), cylinders, strakes, and helicoids.

In Problem **7.6** we will study ruled surfaces and show the following:

> *On a smooth surface M, if α is a geodesic with nonzero normal curvature at each point, then a ribbon can be laid flat along α.*

Note that if a geodesic has zero normal curvature at every point, then (since the extrinsic curvature is equal to the normal curvature) the geodesic is actually extrinsically straight. For such geodesics we do not need the Ribbon Test. The reader can check that the center line of the helicoid is such an extrinsically straight geodesic along which you cannot lay flat a ribbon. (See Problem **1.6**.)

Chapter 4

Tangent Space, Metric, and Directional Derivative

In this chapter we will begin to set up the formal machinery that will allow us to talk about curvature on a surface. We will use this terminology and formalism to obtain an equation for the normal curvature of any curve on the surface. (See Problem **4.7**.) In Chapter 5 we will use this formalism as a part of our intrinsic description of intrinsic curvature. The expression in Problem **4.7** for the normal curvature at a point on a curve depends only on the direction of the curve at that point. Thus it will be the starting point for our investigation of the curvature of the surface in Chapter 6. Note from Chapter 3 we know that the normal curvature is due to the curving of the surface and not due to any intrinsic curving of curves in the surface.

PROBLEM 4.1. The Tangent Space

Go back to Problem **3.1** for the discussion of smooth surfaces and their tangent spaces and normal spaces.

If a curve C intersects a plane Π at a point p, and, if we zoom in on p and find that the portion of the curve in the f.o.v. becomes indistinguishable from a subset of the plane, then we would say that C is ***tangent*** to the plane at p. But clearly this does not mean that C lies in the plane. Thus, in general, for a curve that is tangent to the plane at p, as we zoom in, the portion of the curve in the f.o.v. becomes closer and closer to the plane until it becomes indistinguishable from it. However, when the curve is straight (such as a vector) then, as we zoom in, we see the same picture at all magnifications. (See Figure 4.1.) Which angles we can distinguish depend on the tolerance. With decreasing tolerances we will be able to distinguish smaller and smaller angles. Put this discussion together to show that:

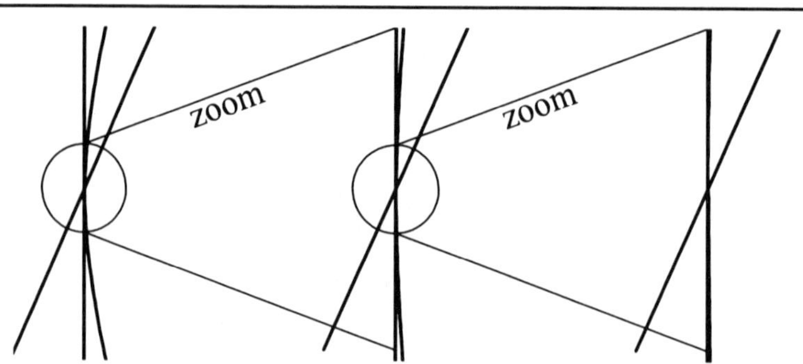

Figure 4.1. Zooming in on two straight lines and a tangent curve

a. *A straight line that is tangent to a plane is contained in the plane.*

Thus, note that tolerances must increase as we zoom in. This is not the same as "zooming in" in computer graphics because the normal computer screen has fixed pixels and thus fixed tolerance. To model on a computer the type of zooming in we are considering here you would need a computer screen capable of decreasing tolerances (or of decreasing the size of the pixels and increasing their number). In addition, you would need to use "vector graphics" (common in drawing programs), not "pixel graphics" (common in paint programs).

Note that a tangent line [or plane] approximates a curve [or surface] in a very different way than 22/7 approximates π. As we zoom in on 22/7 and π, they will become further and further apart. On the contrary, as we zoom in on a point of tangency, the two objects concerned become closer and closer in the f.o.v. and eventually become indistinguishable. It would be useful at this point for the reader to experiment with a favorite function graphing program that is capable of zooming, such as *Analyser**©.

If p *is a point on a smooth surface* M *in* $\mathbb{R}^n$ *and* p *is taken to be the origin of* $\mathbb{R}^n$, *then:*

b. *Show that, for every parametrized curve* $\mathbf{p}(t)$ *that lies in* M *with* $\mathbf{p}(0) = p$, *the velocity vector* $\mathbf{p}'(0)$ *is contained in* T_pM.

[Hint: It is not true in a single f.o.v. that, if A is indistinguishable from B and B is indistinguishable from C, then A is indistinguishable from C. However, you can argue that "being tangent" is a transitive relation, at least in this case.]

c. *Show that every vector lying in T_pM is the velocity vector of some parametrized curve lying in M.*

For Part **c** consider the intersection of M with (n-1)-dimensional subspaces determined by a tangent vector in T_pM and the whole normal space N_pM. Start by looking only at surfaces in $\mathbb{R}^3$ and note that in $\mathbb{R}^3$ the normal space is a normal line. (See Figure 4.2 for a picture of this situation in $\mathbb{R}^3$.)

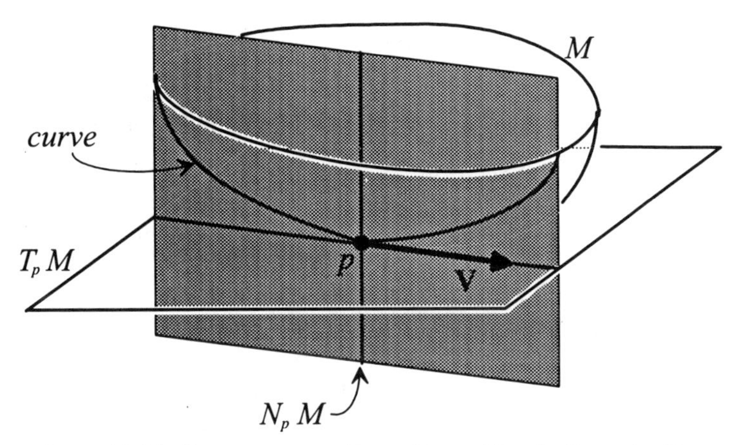

Figure 4.2. Finding a curve, on the surface, with velocity vector V.

We are talking about *parametrized* curves because the velocity of a curve only makes sense if the curve is parametrized. Different parametrizations will, in general, have different speeds but their directions will be the same (or opposite).

The tangent plane is extrinsic since, in general, the tangent vectors do not stay in the surface. But Problem **4.1** allows us to describe a tangent plane intrinsically by saying that the (***intrinsic***) ***tangent plane*** T_pM is the collection of velocity vectors of parametrized curves in M, that pass through p. It is still easier for us to think of these velocity vectors extrinsically, but they also would have intrinsic meaning to a 2-dimensional bug in the surface. In a small neighborhood of p the bug would experience the surface as a flat plane and would have no problem visualizing velocity of curves in that plane. This is the same as our experience in the physical three-dimensional universe. We experience the portion of the universe around us as being a Euclidean space in which we can imagine velocity and other vectors. It does not matter to our imagination what the global geometry of the universe is because for vectors that lie in our immediate space, we can imagine addition and scalar multiplication according to the usual rules.

See Problem **8.2** for further intrinsic descriptions of tangent planes (and tangent spaces).

PROBLEM 4.2. Mean Value Theorem: Curves, Surfaces

In Problem **2.6.b** we pointed out that there was a:

a. ***Mean Value Theorem for Planar Curves.*** *Given two (not equal) points* $\mathbf{p}$ *and* $\mathbf{q}$ *on a differentiable curve in the plane, there is some point* $\mathbf{r}$ *on the curve between* $\mathbf{p}$ *and* $\mathbf{q}$ *such that the tangent vector* $\mathbf{T}_\mathbf{r}$ *at* $\mathbf{r}$ *is parallel to* $\mathbf{p} - \mathbf{q}$. *Prove this theorem.*

[Hint: Look at the line determined by $\mathbf{p}$ and $\mathbf{q}$ and then move this line parallel to itself until it last touches the curve.]

As we pointed out in **2.6.b** this mean value theorem is not true for curves in 3-space. However, we *can* prove:

b. ***Mean Value Theorem for Space Curves.*** *Let* λ *be a smooth curve in* $\mathbb{R}^n$ *with two distinct points a and b and let L be any* $((n-1)$*-dimensional) hyperplane in* $\mathbb{R}^n$ *which contains a and b. Then there is at least one point c on* λ *between a and b such that the line tangent to* λ *at c is parallel to some line in L.*

[Hint: Use the same idea as in Part **a**.]

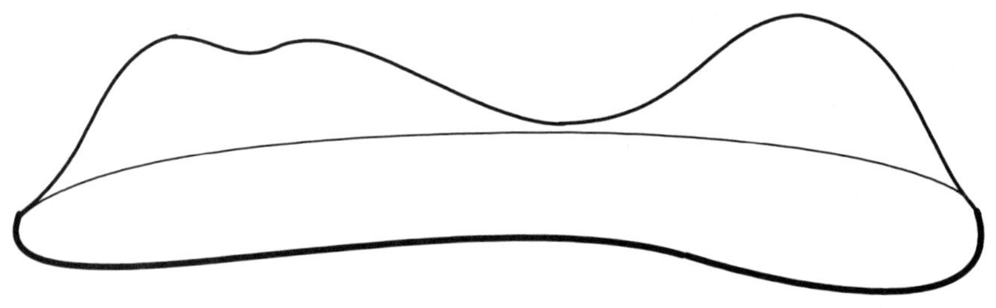

Figure 4.3. Smooth surface with boundary.

There is also an extension of this for surfaces with boundary in 3-space. A ***smooth surface with boundary*** is intuitively a "smooth surface with an edge" and can be defined by specifying that away from the boundary, the surface is smooth as defined above and that the boundary is a smooth curve, and at each point along the boundary the surface has a tangent half-plane (that is, as you zoom in on the point, the surface becomes indistinguishable from a half-plane whose bounding line is the tangent line to the bounding curve). (See Figure 4.3.)

c. ***Mean Value Theorem for Surfaces with Planar Boundary.*** *If M is a differentiable surface in 3-space whose boundary is a planar curve, then some point on the surface has a tangent plane which is parallel to the plane containing the bounding curve.*

[Hint: Use an argument similar to the one used in part **a.**.]

Some more terminology: A surface is said to be ***bounded*** if it is contained in the interior of a finite sphere. A surface is called ***bounding*** if it is the boundary of a volume in 3-space. If a bounded surface has no boundary, then it is called a ***closed surface***. You can extend the above Mean Value Theorems to closed surfaces and prove:

d. *If M is a closed differentiable surface and P is any plane in 3-space, then there are at least two points on the surface whose tangent plane is parallel to P.*

Natural Parametrizations of Curves

If you wish *to find extrinsically a curve on a smooth surface in* $\mathbb{R}^3$ *that has a given tangent vector*, $\mathbf{T}_p$, then you may proceed as in Problem 4.1.b and intersect the surface with the plane determined by $\mathbf{T}_p$ and **n,** the normal to the surface. If you wish *to find intrinsically a curve on a smooth surface that has a given tangent vector*, $\mathbf{T}_p$, then imagine starting at p and proceeding straight in the direction of the tangent vector along a geodesic. The geodesic may be determined by using the ribbon test or the local intrinsic notions of symmetry.

The most *natural intrinsic parametrization of a curve* is by arc-length. Start at some point on the curve and choose a positive direction along the curve. The parameter of a point on the curve is the distance (measured along the curve) from the starting point to that point. Or, if you want a parametrization with constant (nonzero) speed, then let the parameter of a point be the time it takes to go at the constant speed from the starting point to that point; that is, $\gamma(t)$ is the point on the curve that is a distance vt from the starting point $\gamma(0)$, where v is the constant speed.

Alternatively, if we have a smooth curve C in $\mathbb{R}^n$, then we can use any vector $\mathbf{X}$ that is tangent to the curve at p to determine a *natural extrinsic parametrization* of the curve in a neighborhood of p. Pick rectangular coordinates for $\mathbb{R}^n$ so that

$$p = (0,0,0,...,0)$$

and

$$\mathbf{X} = (v,0,0,...,0).$$

Then let

$$g(a,b,c,...,z) = (a,0,0,...,0)$$

be the projection onto the tangent line at p. (See Figure 4.4.) Suppose $g|C$ (the projection restricted to C) is not one-to-one in a neighborhood of p. Then there is a sequence of point pairs $\{a_n,b_n\}$ on C such that $g(a_n) = g(b_n)$, for all n. Let l_n be the line segment joining a_n to b_n. Applying Problem **4.2.b**, there is a point c_n on l_n between a_n and b_n such that a vector tangent to C projects to a point on the tangent line. But then the tangent lines to C cannot be varying continuously. Thus, $g|C$ will be one-to-one in some neighborhood of p. In that neighborhood there is a function

$$\mathbf{x}\colon \mathbb{R} \to C \text{ such that } g(\mathbf{x}(t)) = (vt,0,0,...,0).$$

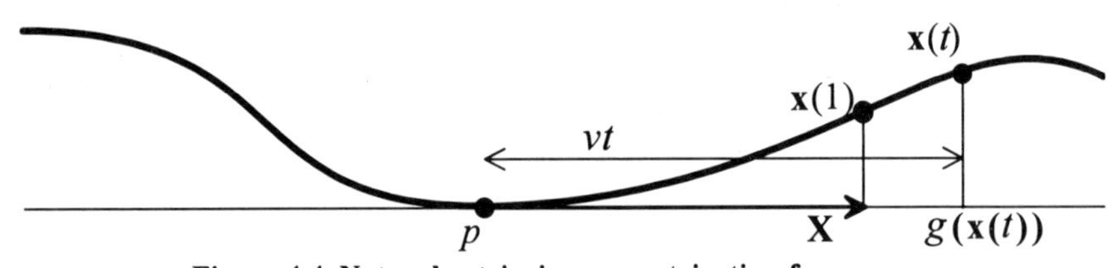

Figure 4.4. Natural extrinsic parametrization for curves.

Note that $\mathbf{x}(t)$ is the intersection of the curve with the $(n-1)$-dimensional subspace that is perpendicular to the tangent line at the point $(vt,0,0,...,0)$. In practice it is usually very difficult, if not impossible, to find an analytic expression for this function. This $\mathbf{x}$ gives a parametrization for a neighborhood of p in C and $\mathbf{x}'(0) = \mathbf{X}$.

Note that every smooth surface has an extrinsic Monge patch $\mathbf{x}$ (Problem **1.9**) that is C^1 (Problem **3.1.e**). If

$$\mathbf{X}_p = X^1\mathbf{x}_1 + X^2\mathbf{x}_2,$$

then

$$\gamma(t) = \mathbf{x}(X^1 t, X^2 t)$$

is a curve on M with velocity vector $\mathbf{X}_p$. In $\mathbb{R}^3$, this curve is a parametrized version of the intersection of the surface with the plane spanned by $\mathbf{X}_p$ and $\mathbf{n}$.

Problem 4.3. Riemannian Metric

If M is a smooth surface in $\mathbb{R}^n$, then the (induced) ***Riemannian metric*** (or ***first fundamental form***) at $p \in M$ is defined by

$$\langle \mathbf{X}, \mathbf{Y} \rangle = |\mathbf{X}|\,|\mathbf{Y}| \cos \theta_{\mathbf{XY}},$$

for $\mathbf{X}$ and $\mathbf{Y}$ vectors in T_pM, where $\theta_{\mathbf{XY}}$ is the angle between $\mathbf{X}$ and $\mathbf{Y}$. This is an intrinsic definition, because vectors in T_pM can be intrinsically described as velocity vectors, and thus, their lengths are speeds and the angle θ between the directions of two vectors is a quantity that can be intrinsically measured. The reason for the term "*metric*" is that it will allow us (see Problem **4.5**) to express lengths of curves and areas of regions. The fact that the Riemannian metric is bilinear (**4.3.a**) will allow us later to use local coordinates in powerful ways.

It is precisely **4.3.a** that allows us to express the Riemannian metric in local coordinates. Thus it would lead to circular reasoning if we used local coordinates to prove **4.3.a**. We need to know that the important properties in **4.3.a** are geometric properties that follow directly from the geometric definition and hold regardless of the coordinate system.

a. *Show that the Riemannian metric is:*

symmetric, *that is,* $\langle \mathbf{X}, \mathbf{Y} \rangle = \langle \mathbf{Y}, \mathbf{X} \rangle$;

bilinear, *that is,* $a\langle \mathbf{X}, \mathbf{Y} \rangle = \langle a\mathbf{X}, \mathbf{Y} \rangle = \langle \mathbf{X}, a\mathbf{Y} \rangle$, *for* $a \in \mathbb{R}$,
$$\langle \mathbf{X}, \mathbf{Y} + \mathbf{Z} \rangle = \langle \mathbf{X}, \mathbf{Y} \rangle + \langle \mathbf{X}, \mathbf{Z} \rangle$$

positive definite, $\langle \mathbf{X}, \mathbf{X} \rangle$ *is positive, if* $\mathbf{X} \neq \mathbf{0}$.

[Hint: If $a < 0$ then

$$\theta_{a\mathbf{XY}} = \theta_{\mathbf{XY}} + \pi.$$

Be sure you see why and take this into account when showing the bilinearity. To show that

$$\langle \mathbf{X}, \mathbf{Y} + \mathbf{Z} \rangle = \langle \mathbf{X}, \mathbf{Y} \rangle + \langle \mathbf{X}, \mathbf{Z} \rangle$$

draw a picture and look at projections.]

b. *If* $\mathbf{X}_1, \mathbf{X}_2$ *is an orthonormal basis for* T_pM,

$$\text{(that is, } \langle \mathbf{X}_1, \mathbf{X}_2 \rangle = 0 \text{ and } \langle \mathbf{X}_1, \mathbf{X}_1 \rangle = 1 = \langle \mathbf{X}_2, \mathbf{X}_2 \rangle)$$

and

$$\mathbf{A} = a_1\mathbf{X}_1 + a_2\mathbf{X}_2 \text{ and } \mathbf{B} = b_1\mathbf{X}_1 + b_2\mathbf{X}_2,$$

then show that

$$\langle \mathbf{A}, \mathbf{B} \rangle = a_1b_1 + a_2b_2.$$

c. *If* $\mathbf{X}_1, \mathbf{X}_2$ *is an arbitrary basis for* T_pM, *and*

$$\mathbf{A} = a_1\mathbf{X}_1 + a_2\mathbf{X}_2 \text{ and } \mathbf{B} = b_1\mathbf{X}_1 + b_2\mathbf{X}_2,$$

then show that

$$\langle \mathbf{A}, \mathbf{B} \rangle = \begin{pmatrix} a_1 & a_2 \end{pmatrix} \begin{pmatrix} g_{11} & g_{12} \\ g_{21} & g_{22} \end{pmatrix} \begin{pmatrix} b_1 \\ b_2 \end{pmatrix},$$

where

$$g_{ij} = \langle \mathbf{X}_i, \mathbf{X}_j \rangle.$$

Note that $\langle \mathbf{A}, \mathbf{B} \rangle$ is equal to the inner product $a_1b_1 + a_2b_2$ only when the basis $\mathbf{X}_1, \mathbf{X}_2$ is orthonormal and thus the matrix (g_{ij}) is the identity matrix. It is always possible to pick a coordinate system that is orthonormal at a particular point, but it is *not* possible to find a coordinate system that is orthonormal at all points unless the surface is developable from the plane (locally isometric to the plane).

In most texts the matrix $g = (g_{ij})$ is called the *first fundamental form* or *metric* (with respect to the basis $\mathbf{X}_1$, $\mathbf{X}_2$). Clearly, with different bases the matrices g will, in general, be different.

Note that the above definition of $\langle \mathbf{X},\mathbf{Y} \rangle$ makes sense in any dimension. In particular, $\langle \mathbf{X},\mathbf{Y} \rangle$ depends only on a plane (2-dimensional subspace) that contains **X** and **Y**. If **X** and **Y** are linearly independent (do not lie in the same line), then there is a unique plane, sp(**X**,**Y**), which is determined (spanned) by **X** and **Y**. If we look at that plane, we see that

$$\langle \mathbf{X},\mathbf{Y} \rangle = |\mathbf{X}|\ (|\mathbf{Y}| \cos \theta),$$

where $|\mathbf{Y}| \cos \theta$ is the length of the projection of **Y** onto **X** with a negative sign if **X** and the projection of **Y** point in opposite directions. (See Figure 4.5.)

Note that the picture in Figure 4.5 holds on the tangent plane (which has Euclidean geometry) and, in general, does not hold on a surface that does not have Euclidean geometry.

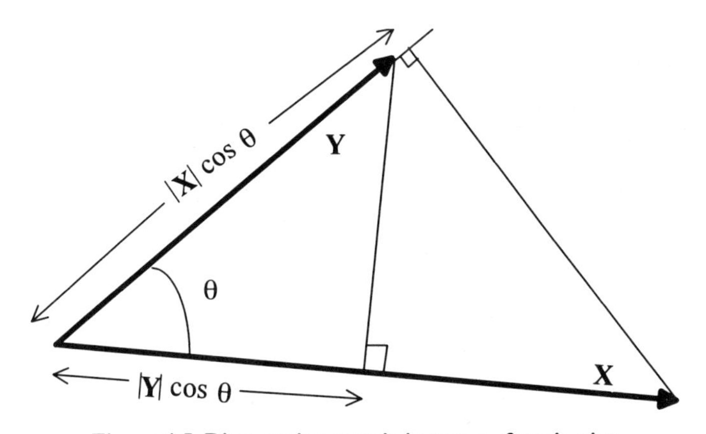

Figure 4.5. Riemannian metric in terms of projections.

It is usually best when working with the Riemannian metric to avoid manipulating the $\cos \theta$ because one is likely to get involved in the unnecessary complication of trigonometric identities. Usually, it is enough to look at the Riemannian metric geometrically in terms of projections as in Figure 4.5 or to use (after you have proved it!) the bilinearity of the metric.

It is particularly important to start with a coordinate free description of the Riemannian metric because, even if the surface is sitting in $\mathbb{R}^n$ with a coordinate system, there will not, in general, be a natural coordinate system on a tangent plane, T_pM, to the surface. In different settings, we may choose a coordinate system on the tangent plane, but different settings will naturally lead to different coordinate systems. For example, if we want to focus on a particular curve through p, then we may want the basis vectors for T_pM to be the unit tangent vector and the unit intrinsic normal to the curve at p. If we wish to focus on the curvature of the surface then it is often convenient to choose the principal directions as the coordinate directions (see discussion of curvature and principal directions in Chapter 6).

Note that $\langle \mathbf{X},\mathbf{Y} \rangle$ can be interpreted as an inner product in any space in which it sits with respect to an orthonormal coordinate system. For example, if **X** and **Y** are tangent vectors at p on a surface M, which is in $\mathbb{R}^n$, then (as in Problem **4.3**) $\langle \mathbf{X},\mathbf{Y} \rangle$ is the inner product with respect to any two-dimensional orthonormal coordinate system on T_pM, but it is also the inner product with respect to any orthonormal coordinate system in $\mathbb{R}^n$.

Riemannian Metric in Local Coordinates on a Sphere

We use the sphere to develop an example of expressing the Riemannian metric in terms of (extrinsically defined) local coordinates.

A point on the sphere of radius r can be expressed in terms of two coordinates, θ, ϕ, by this formula (see Figure 4.6):

$$\mathbf{x}(\theta,\phi) = (r\cos\theta\sin\phi, r\sin\theta\sin\phi, r\cos\phi).$$

Note that $\mathbf{x}$ is a map from $\mathbb{R}^2$ into $\mathbb{R}^3$. At the point $p = \mathbf{x}(\theta,\phi)$ the coordinate curves are

$$\lambda(t) = \mathbf{x}(t,\phi) \text{ and } \gamma(t) = \mathbf{x}(\theta,t).$$

We can obtain a basis for the tangent space at p by using the velocity vectors of these curves:

$$\frac{d}{dt}\lambda(t)_{t=\theta} = \frac{\partial}{\partial t}\mathbf{x}(t,\phi)_{t=\theta} \equiv \mathbf{x}_1(\theta,\phi) = (-r\sin\theta\sin\phi, r\cos\theta\sin\phi, 0),$$

$$\frac{d}{dt}\gamma(t)_{t=\theta} = \frac{\partial}{\partial t}\mathbf{x}(\theta,t)_{t=\phi} \equiv \mathbf{x}_2(\theta,\phi) = (r\cos\theta\cos\phi, r\sin\theta\cos\phi, -r\sin\phi).$$

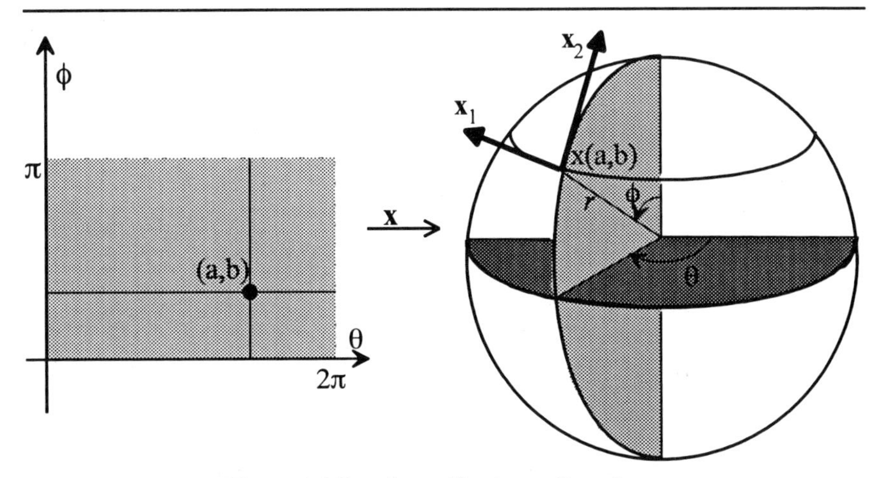

Figure 4.6. Local coordinates on the sphere.

We can now express the Riemannian metric in terms of these local coordinates using Problem **4.3.c**:

$$g_{11}(\theta,\phi) = \langle \mathbf{x}_1,\mathbf{x}_1\rangle = |\mathbf{x}_1|^2 = r^2\sin^2\phi,$$
$$g_{22}(\theta,\phi) = \langle \mathbf{x}_2,\mathbf{x}_2\rangle = |\mathbf{x}_2|^2 = r^2,$$
$$g_{12}(\theta,\phi) = \langle \mathbf{x}_1,\mathbf{x}_2\rangle = 0 = \langle \mathbf{x}_2,\mathbf{x}_1\rangle = g_{21}(\theta,\phi),$$

and thus the matrix of the Riemannian metric is:

$$g_{ij} = \begin{pmatrix} r^2\sin^2\phi & 0 \\ 0 & r^2 \end{pmatrix}.$$

Note that this shows that $\{\mathbf{x}_1, \mathbf{x}_2\}$ is an orthonormal basis only on the equator of a unit sphere.

Riemannian Metric in Local Coordinates on a Strake

We use the strake to develop another example of expressing the Riemannian metric in terms of (extrinsically defined) local coordinates. (See Figure 4.7.)

A point on the strake can be expressed in terms of two coordinates, r, θ, by this formula:

$$\mathbf{x}(\theta,r) = (r\cos\theta,\ r\sin\theta,\ k\theta).$$

Note that $\mathbf{x}$ is a map from $\mathbb{R}^2$ into $\mathbb{R}^3$.

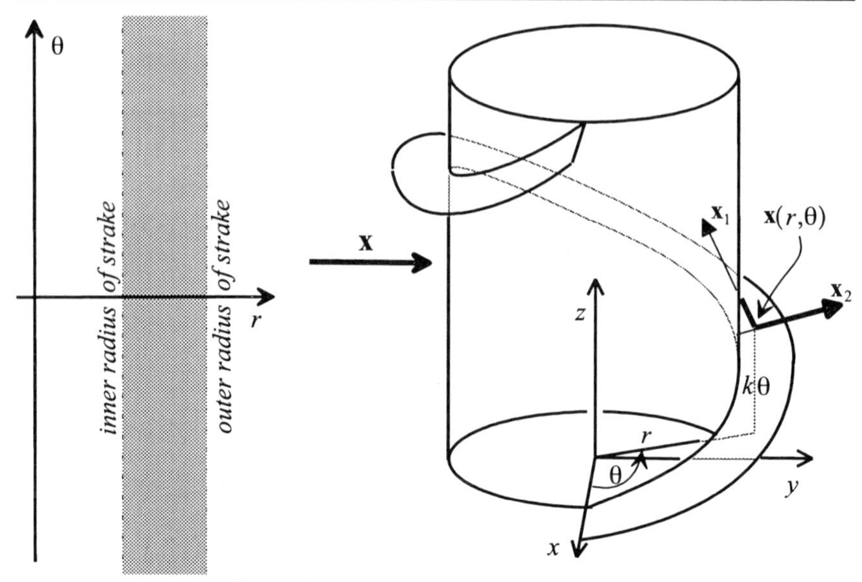

Figure 4.7. Local coordinates on a strake.

At the point $p = \mathbf{x}(\theta,r)$ the coordinate curves are $\lambda(t) = \mathbf{x}(t,r)$ and $\gamma(t) = \mathbf{x}(\theta,t)$. We can obtain a basis for the tangent space at p by using the velocity vectors of these curves:

$$\frac{d}{dt}\lambda(t)_{t=\theta} = \frac{\partial}{\partial t}\mathbf{x}(t,r)_{t=\theta} \equiv \mathbf{x}_1(\theta,r) = (-r\sin\theta, r\cos\theta, k)$$

$$\frac{d}{dt}\gamma(t)_{t=r} = \frac{\partial}{\partial t}\mathbf{x}(\theta,t)_{t=r} \equiv \mathbf{x}_2(\theta,r) = (\cos\theta, \sin\theta, 0).$$

Now we can express the Riemannian metric in terms of these coordinates using Problem **4.3.c**:

$$g_{11}(\theta,r) = \langle \mathbf{x}_1,\mathbf{x}_1\rangle = |\mathbf{x}_1|^2 = r^2 + k^2,$$

$$g_{22}(\theta,r) = \langle \mathbf{x}_2,\mathbf{x}_2\rangle = |\mathbf{x}_2|^2 = 1,$$

$$g_{12}(\theta,r) = \langle \mathbf{x}_1,\mathbf{x}_2\rangle = 0 = \langle \mathbf{x}_2,\mathbf{x}_1\rangle = g_{21}(\theta,r).$$

Thus the matrix of the Riemannian metric in these coordinates is:

$$[g_{ij}] = \begin{pmatrix} r^2 + k^2 & 0 \\ 0 & 1 \end{pmatrix}.$$

Note that this is an orthogonal coordinate system and, although it is impossible to make it orthonormal at every point, it is possible that at some point it will be orthonormal (that is, when $r^2 + k^2 = 1$).

Thus if

$$\mathbf{X}_p = a\mathbf{x}_1 + b\mathbf{x}_2 \text{ and } \mathbf{Y}_p = c\mathbf{x}_1 + d\mathbf{x}_2$$

are two tangent vectors at p, then we can write

$$\langle \mathbf{X}_p, \mathbf{Y}_p \rangle = \begin{pmatrix} a & b \end{pmatrix} \begin{pmatrix} r^2 + k^2 & 0 \\ 0 & 1 \end{pmatrix} \begin{pmatrix} c \\ d \end{pmatrix} = ac(r^2 + k^2) + bd.$$

Problem 4.4. Vectors in Extrinsic Local Coordinates

If M is a smooth surface in $\mathbb{R}^n$ and $\mathbf{x}: U \to M$ is a one-to-one function defined on an open region U in the (u^1,u^2)-plane with values in M, then we call $\mathbf{x}$ a **C¹ *coordinate patch*** (or ***C¹ local coordinates***) ***for*** M if, as a function from $\mathbb{R}^2$ to $\mathbb{R}^n$, $\mathbf{x}$ is C^1 (that is, $\mathbf{x}$ is differentiable and the partial derivatives are continuous), and the vectors $\mathbf{x}_1(a,b)$, $\mathbf{x}_2(a,b)$ are linearly independent for each (a,b) in U, where, if $p = \mathbf{x}(a,b)$, then $\mathbf{x}(a,u^2)$ and $\mathbf{x}(u^1,b)$ are curves on M, and their velocity vectors at p are

$$\mathbf{x}_1(a,b) = \frac{\partial}{\partial u^1}\mathbf{x}(u^1,b)_{u^1=a};\ \mathbf{x}_2(a,b) = \frac{\partial}{\partial u^2}\mathbf{x}(a,u^2)_{u^2=b}.$$

In $\mathbb{R}^3$, the normal to the surface at $p = \mathbf{x}(a,b)$ is perpendicular to both of the tangent vectors $\mathbf{x}_1(a,b)$ and $\mathbf{x}_2(a,b)$ and thus the normal can be expressed as the unit vector:

$$\mathbf{n}(a,b) = \frac{\mathbf{x}_1(a,b) \times \mathbf{x}_2(a,b)}{|\mathbf{x}_1(a,b) \times \mathbf{x}_2(a,b)|}.$$

Note that at every point on the surface there are two possible normals (in opposite directions to each other), and the above expression picks one of these continuously over the coordinate patch. However, most surfaces cannot be covered by a single coordinate patch. Spheres and cylinders need at least two coordinate patches, yet it is possible to make a continuous selection of normal over the whole surface—such surfaces are said to be ***orientable***. But on some surfaces (for example, a Moebius band), it is not possible to continuously pick a normal at every point—such a surface is said to be a ***non-orientable***. For further discussion of orientable and non-orientable surfaces see Jeff Weeks' delightful book, *The Shape of Space*, [**DG**: Weeks].

a. *For each of the surfaces cylinder, cone, sphere, strake, surfaces of revolution, and the graph of a smooth function* $z = f(x, y)$, *use the (extrinsically defined) local coordinates from Chapter 1 and find in each case an expression for* $\mathbf{x}_1$ *and* $\mathbf{x}_2$. *Check that the local coordinates are a* C^1 *coordinate patch.*

Since $\mathbf{x}_1(a,b)$, $\mathbf{x}_2(a,b)$ are linearly independent, they form a basis for the tangent (vector) space T_pM. Therefore a vector $\mathbf{X}_p$ in T_pM can be expressed as

$$\mathbf{X}_p = X^1\mathbf{x}_1(a,b) + X^2\mathbf{x}_2(a,b) = \Sigma\, X^i\mathbf{x}_i(a,b),$$

where X^1, X^2 are real numbers. Our convention is to use superscripts on coefficients and subscripts on basis vectors.

If we have a vector field defined on M [that is, a vector-valued function

$$\mathbf{X}_p = \mathbf{X}(p) = \mathbf{X}(\mathbf{x}(u^1,u^2)),$$

where $\mathbf{X}(\mathbf{x}(u^1,u^2))$ is in the tangent space at $p = \mathbf{x}(u^1,u^2)$], then we can write

$$\begin{aligned}\mathbf{X}(p) &= \mathbf{X}(\mathbf{x}(u^1,u^2)) = \\ &= X^1(u^1,u^2)\mathbf{x}_1(u^1,u^2) + X^2(u^1,u^2)\mathbf{x}_2(u^1,u^2) \\ &= \Sigma\, X^i(u^1,u^2)\mathbf{x}_i(u^1,u^2).\end{aligned}$$

But usually we implicitly assume the coordinate variables and simply write

$$\mathbf{X}_p = X^1\mathbf{x}_1 + X^2\mathbf{x}_2 = \Sigma\, X^i\mathbf{x}_i\,.$$

In many texts this expression is simplified even further by using the so-called "summation convention." In the summation convention there is an implied summation over all repeated indices with one as a subscript and one as a superscript: $\mathbf{X}_p = X^i\mathbf{x}_i$.

Now we can use Problem **4.3.c** to express the Riemannian metric applied to the vector fields

$$\mathbf{X}_p = \Sigma\, X^i \mathbf{x}_i \text{ and } \mathbf{Y}_p = \Sigma\, Y^i \mathbf{x}_j$$

in local coordinates as

$$\langle \mathbf{X}_p, \mathbf{Y}_p \rangle = \begin{pmatrix} X^1 & X^2 \end{pmatrix} \begin{pmatrix} g_{11} & g_{12} \\ g_{21} & g_{22} \end{pmatrix} \begin{pmatrix} Y^1 \\ Y^2 \end{pmatrix} = \Sigma\, X^i g_{ij} Y^j,$$

where $g_{ij} = \langle \mathbf{X}_i, \mathbf{X}_j \rangle$.

b. *For each of the coordinate patches in part **a**, determine the matrix of the Riemannian metric.*

Warning: Some texts call (g_{ij}) simply the Riemannian metric, but this only makes sense if a local coordinate system is being assumed (either implicitly or explicitly). When you change the local coordinates, then the matrix (g_{ij}) usually changes.

PROBLEM 4.5. Measuring Using the Riemannian Metric

To find the distance (arclength) along any path γ we may integrate the speed $|\gamma'|$ along the path. Thus, the arclength between $\gamma(a)$ and $\gamma(b)$ is equal to

$$\int_a^b |\gamma'(t)|\, dt = \int_a^b \sqrt{\langle \gamma'(t), \gamma'(t) \rangle}\, dt.$$

Note that this is an ordinary integral studied in first-year calculus. So, in principle (or, theoretically), once the Riemannian metric is known we can determine the arclength of any path and thus (using geodesics segments) the distance along the surface between points. I say "in principle" because, in general, the presence of the radical makes solving the integral very difficult. There is only a small special class of curves for which it is possible to evaluate the integral exactly. You can find examples of these curves in the exercises at the end of the arclength section of standard calculus texts. In a recent article [**Z**: Pottmann, p.183], there is a description of all planar curves with parametrizations, which are rational functions and for which the arclength integral is solvable as a rational function. For most curves the arclength can be calculated only approximately.

If $\mathbf{x}(u^1,u^2)$ gives local coordinates for a region R on the surface, then, in a f.o.v. in which the surface is indistinguishable from the tangent plane, a change of coordinate from

$$u^1 \text{ to } u^1+\Delta u^1 \text{ [or, from } u^2 \text{ to } u^2+\Delta u^2\text{]}$$

will produce a segment on the surface of length

$$\int_{u^1}^{u^1+\Delta u^1} |\mathbf{x}_1(t,u^2)|\, dt \quad [\text{or}, \int_{u^2}^{u^2+\Delta u^2} |\mathbf{x}_2(u^1,t)|\, dt\,] .$$

If we let Δu^1 and Δu^2 be small enough, then these segments on the surface will become indistinguishable from straight line segments and we get a small parallelogram on the surface. (See Figure 4.8.)

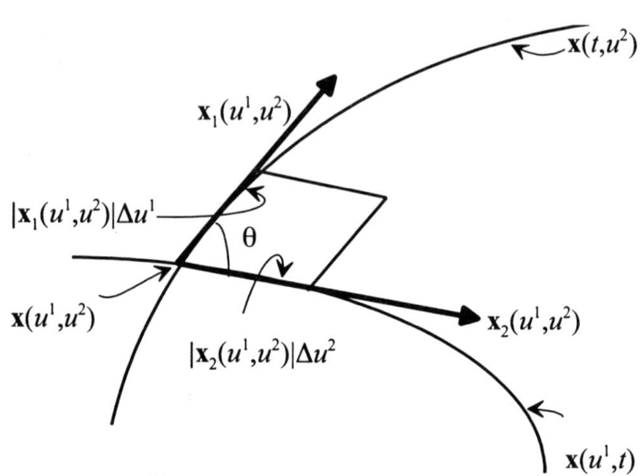

Figure 4.8. Calculating surface area.

a. *Show that this parallelogram has area*:

$$(|\mathbf{x}_1(u^1,u^2)|\Delta u^1)(|\mathbf{x}_2(u^1,u^2)|\Delta u^2)(\sin\theta) = \sqrt{\det g(u^1,u^2)}\;\Delta u^1 \Delta u^2 ,$$

where $g(u^1,u^2) = [g_{ij}(u^1,u^2)]$ *is the matrix of the Riemannian metric with respect to the coordinates* $\mathbf{x}(u^1,u^2)$.

[Hint: Use the definition of the Riemannian metric to find an expression for $\sin\theta$.]

We can now let Δu^1 and Δu^2 go to zero and integrate over the region R to get the following expression for the area of R:

$$\iint_R \sqrt{\det g(u^1,u^2)}\; du^1 du^2.$$

b. *Use the above expression to find the surface area of an intrinsic circular disk of (intrinsic) radius r on a sphere of radius R. Compare this to the area of a circular disk of the same radius on a plane in the case of* $r = 1$ km *and* $R = 6360$ km *(the approximate radius of the earth).*

[Hint: Use spherical coordinates from Problem **4.4**, and choose the North pole as the center of the circle. You will need to evaluate the integral as an improper integral (*Why?*).]

c. *On a cone with cone angle* α, *use the above expression to find the surface area of an intrinsic circular disk of (intrinsic) radius r with the (intrinsic) center of the circle at the cone point.*

[Hint: Similar to Part **c**.]

d. *Find the area of one turn of the strake, and compare this area to the area of the annular strip that approximates it. Use the dimensions given in Problem* **2.5**.

[Hint: Use Problems **2.5** and **4.4.** This is a somewhat tedious computation, but it can be done, especially if you use an integral table or computer algebra system to evaluate the integrals that appear.]

Directional Derivatives

If f is a function (real-valued or vector-valued) defined on the smooth surface M, and $\mathbf{X}_p$ is a tangent vector in the tangent plan T_pM, then we can define the (extrinsic) ***directional derivative of f with respect to*** $\mathbf{X}_p$ by

$$\mathbf{X}_p f = \frac{d}{dt} f(\gamma(t))_{t=0} = \lim_{h \to 0} \frac{f(\gamma(h)) - f(\gamma(0))}{h},$$

where $\gamma(t)$ is any parametrized curve with $\gamma(0) = p$ and $\gamma'(0) = \mathbf{X}_p$. (Such a γ exists by Problem **4.1**.) Thus $\mathbf{X}_p f$ is the rate of change of f as one travels along the curve γ at the point p. The function f is said to be ***differentiable at*** p if $\mathbf{X}_p f$ exists independent of the choice of the curve γ (such that $\gamma(0) = p$ and $\gamma'(0) = \mathbf{X}_p$) for each tangent vector $\mathbf{X}_p$ in T_pM. Note that in order for f to be differentiable, it must be a function defined on some neighborhood of p, not just on γ or just at p. Below we will show that

$\mathbf{X}_p f$ is independent of the choice of curve whenever there is a local coordinate patch $\mathbf{x}$ such that $f \circ \mathbf{x}$ is differentiable.

Many texts use the terminology *directional derivative in the direction of* **X**. That terminology originated because at first people only considered directional derivatives with respect to unit vectors and thus did not consider the lengths of the vectors. We do not follow that terminology because, in a literal sense, we are not differentiating *in the direction of a vector* because the directional derivative depends on both the direction and the length of the vector.

If our surface M is the plane, then the rate of change of f along a curve γ is the derivative of $f(\gamma(t))$ with respect to t. But by the chain rule, at $t = 0$, this is

$$f'(\gamma(0))\,\gamma'(0) = f'(\gamma(0))\,\mathbf{X}_p.$$

This result is one of the motivations for the notation. In addition, we will show in Problem **4.8** that $\mathbf{X}_p f$ depends linearly on $\mathbf{X}_p$, and thus the notation makes it convenient to express this linearity as

$$(\mathbf{X}_p + (a\mathbf{Y}_p))f = \mathbf{X}_p f + a(\mathbf{Y}_p f).$$

Examples using the (global) coordinates in $\mathbb{R}^n$:

1. If M is the sphere

$$x^2 + y^2 + z^2 = 4$$

in 3-space and

$$f(x,y,z) = 5x^2 + y + z$$

and $\mathbf{X}_p$ is the tangent vector at $p = (2,0,0)$, which is parallel to $(0,1,0)$, then pick

$$\gamma(t) = (2\cos t/2,\ 2\sin t/2,\ 0).$$

Note that $\gamma'(0) = \mathbf{X}_p$, then

$$f(\gamma(t)) = (5)(4\cos^2 t/2) + (2\sin t/2) + (0)$$

and

$$\mathbf{X}_p f = (5)(4)(2)(\cos 0)(-\sin 0)(1/2) + (2)(\cos 0)(1/2) + (0) = 1.$$

2. For the same M, $\mathbf{X}_p$, p, we can also differentiate the vector-valued function

$$\mathbf{n}(x,y,z) = \tfrac{1}{2}(-x,-y,-z),$$

which gives the unit (inward) normal to M and thus

$\mathbf{n}(\gamma(t)) = \tfrac{1}{2}(-2 \cos t/2, -2 \sin t/2, 0)$

and

$$\mathbf{X}_p\mathbf{n} = (-\tfrac{1}{2} \sin 0, \tfrac{1}{2} \cos 0, 0) = (0,\tfrac{1}{2},0) = \tfrac{1}{2}\, \mathbf{X}_p.$$

Note that in $\mathbf{X}_p\mathbf{n}$, the $\mathbf{X}_p$ is a tangent vector at p but $\mathbf{n}$ is a vector-valued function. (If it were not a function, you would not be able to differentiate it!)

Example in local coordinates:

Let us redo Example 2, above, in local coordinates:

2. If M is the sphere

$$x^2 + y^2 + z^2 = 4$$

in 3-space and $\mathbf{X}_p$ is the tangent vector at $p = (2,0,0)$, which is parallel to $(0,1,0)$, then pick

$$\gamma(t) = (2 \cos t/2, 2 \sin t/2, 0).$$

Then we can differentiate the vector-valued function

$$\mathbf{n}(x,y,z) = \tfrac{1}{2}(-x,-y,-z),$$

which gives the unit (inward) normal to M and thus

$\mathbf{n}(\gamma(t)) = \tfrac{1}{2}(-2 \cos t/2, -2 \sin t/2, 0)$

and

$$\mathbf{X}_p\mathbf{n} = (-\tfrac{1}{2} \sin 0, \tfrac{1}{2} \cos 0, 0) = (0,\tfrac{1}{2},0) = \tfrac{1}{2}\, \mathbf{X}_p.$$

Let

$$\mathbf{x}(\theta,\phi) = (2 \cos\theta \cos\phi, 2 \sin\theta \cos\phi, 2 \sin\phi)$$

be the local coordinate system on the sphere M. Then the point

$$p = (2,0,0) = \mathbf{x}(0,0)$$

and at this point:

$$\mathbf{x}_1 = \mathbf{x}_1(0,0) = \frac{\partial}{\partial \theta}\mathbf{x}(\theta,0)_{\theta=0} =$$
$$= (-2\sin\theta\cos 0, 2\cos\theta\cos 0, 2\sin 0)_{\theta=0} = (0,2,0),$$

and

$$\mathbf{x}_2 = \mathbf{x}_2(0,0) = \frac{\partial}{\partial \phi}\mathbf{x}(0,\phi)_{\phi=0} =$$
$$= (-2\cos 0\sin\phi, -2\sin 0\sin\phi, 2\cos\phi)_{\phi=0} = (0,0,2).$$

Now

$$\mathrm{X}_p = \tfrac{1}{2}\,\mathbf{x}_1 + 0\,\mathbf{x}_2\,,\ \text{so } X^1 = \tfrac{1}{2} \text{ and } X^2 = 0.$$

Then

$$(\mathbf{n}\circ\mathbf{x})(\theta,\phi) = (-\cos\theta\cos\phi, -\sin\theta\cos\phi, -\sin\phi)$$

and we can write $\gamma(t) = \mathbf{x}(t/2,0)$ and at $p = (0,0)$:

$$\begin{aligned}
\mathrm{X}_p\mathbf{n} &= \frac{d}{dt}\mathbf{n}(\gamma(t))_{t=0} = \\
&= \frac{d}{dt}(\mathbf{n}\circ\mathbf{x})(t/2,0)_{t=0} = \\
&= \frac{\partial(\mathbf{n}\circ\mathbf{x})}{\partial\theta}\left(\frac{1}{2}\right) + \frac{\partial(\mathbf{n}\circ\mathbf{x})}{\partial\phi}(0) = \\
&= X^1\mathbf{x}_1\mathbf{n} + X^2\mathbf{x}_2\mathbf{n} = \\
&= \tfrac{1}{2}(0,1,0) + 0 \\
&= \tfrac{1}{2}\,\mathrm{X}_p.
\end{aligned}$$

Directional Derivative in Local Coordinates

More generally, if f is a function (vector-valued or real-valued) defined on M such that $f\circ\mathbf{x}$ is C^1 (continuously differentiable), then the directional derivatives in the directions of $\mathbf{x}_1$ and $\mathbf{x}_2$ are:

$$\mathbf{x}_1 f = \frac{d}{du^1}f(\mathbf{x}(u^1,b))_{u^1=a} = \frac{\partial(f\circ\mathbf{x})}{\partial u^1}(a,b)$$

and

$$\mathbf{x}_2 f = \frac{d}{du^2}f(\mathbf{x}(a,u^2))_{u^2=b} = \frac{\partial(f\circ\mathbf{x})}{\partial u^2}(a,b).$$

Now we can express $\mathbf{X}_p f$ in terms of these coordinates. Let $\alpha(t)$ be a curve on M with

$$\alpha(0) = p \quad \text{and} \quad \dot{\alpha}(0) = \mathbf{X}_p.$$

Then, near p (or when t is near 0) we may write

$$\alpha(t) = \mathbf{x}(\alpha^1(t), \alpha^2(t)).$$

Here $\alpha^1(t), \alpha^2(t)$ are the coordinates of $\alpha(t)$. In this context $\mathbf{x}(u^1, u^2)$ gives the location of the point with coordinates u^1, u^2. We can then write the velocity of α as:

$$\begin{aligned}
\dot{\alpha}(0) &= \frac{d}{dt}\mathbf{x}(\alpha^1(t), \alpha^2(t))_{t=0} = \\
&= (\mathbf{x}_1)\left(\frac{d\alpha^1}{dt}\right)_{t=0} + (\mathbf{x}_2)\left(\frac{d\alpha^2}{dt}\right)_{t=0} = \\
&= X^1\mathbf{x}_1 + X^2\mathbf{x}_2.
\end{aligned}$$

Thus

$$\left(\frac{d\alpha^1}{dt}\right)_{t=0} = X^1 \text{ and } \left(\frac{d\alpha^2}{dt}\right)_{t=0} = X^2.$$

Therefore,

$$\begin{aligned}
\mathbf{X}_p f = \frac{d}{dt}f(\alpha(t))_{t=0} &= \frac{d}{dt}f(\,\mathbf{x}(\alpha^1(t), \alpha^2(t))\,)_{t=0} = \\
&= \frac{d}{dt}(f \circ \mathbf{x})(\alpha^1(t), \alpha^2(t))_{t=0} = \\
&= \frac{\partial(f\circ\mathbf{x})}{\partial u^1}\left(\frac{d\alpha^1}{dt}\right)_{t=0} + \frac{\partial(f\circ\mathbf{x})}{\partial u^2}\left(\frac{d\alpha^2}{dt}\right)_{t=0} = \\
&= X^1\mathbf{x}_1 f + X^2\mathbf{x}_2 f.
\end{aligned}$$

Thus we have shown:

If $f \circ \mathbf{x}$ is C^1, then f is differentiable (that is, the directional derivatives do not depend on the choice of curve.)

Problem 4.6. Differentiating a Metric

Local coordinates are not necessary in any of these parts.

a. *Show that if* **X** *and* **Y** *are differentiable vector-valued functions defined on a curve C with parametrization* $\gamma(t)$, *then*

$$\frac{d}{dt}\langle \mathbf{X}(\gamma(t)), \mathbf{Y}(\gamma(t))\rangle =$$
$$= \left\langle \frac{d}{dt}\mathbf{X}(\gamma(t)), \mathbf{Y}(\gamma(t)) \right\rangle + \left\langle \mathbf{X}(\gamma(t)), \frac{d}{dt}\mathbf{Y}(\gamma(t)) \right\rangle.$$

[Hint: $\langle \mathbf{X}(\gamma(t)),\mathbf{Y}(\gamma(t))\rangle$ is a continuous (*Why?*) real-valued function of a real variable, so you can apply the definition of derivative from first-semester calculus. Avoid expressions where "cos θ" shows.]

b. *Show that if* **X** *and* **Y** *are defined on a neighborhood of p in M, and* $\mathbf{Z}_p$ *is a vector in* T_pM, *then*

$$\mathbf{Z}_p\langle \mathbf{X},\mathbf{Y}\rangle = \langle \mathbf{Z}_p\mathbf{X},\mathbf{Y}(p)\rangle + \langle \mathbf{X}(p),\mathbf{Z}_p\mathbf{Y}\rangle.$$

[Hint: Use Part **a**.]

c. *If* **X** *and* **Y** *are defined (and differentiable) on a neighborhood of p in M and are everywhere perpendicular, then show that:*

$$\langle \mathbf{Z}_p\mathbf{X}, \mathbf{Y}\rangle = -\langle \mathbf{X}, \mathbf{Z}_p\mathbf{Y}\rangle.$$

[Hint: Use Part **b**.]

Problem 4.7. Expressing Normal Curvature

In Chapter 3 we defined the normal curvature of a curve C on a surface M in $\mathbb{R}^3$ to be the projection of the (extrinsic) curvature vector onto the normal line. Now, we express the normal curvature in local coordinates and prepare for the second fundamental form, which will be an important tool in studying the curvature of the surface.

a. *Let C be a smooth curve on a smooth surface M in* $\mathbb{R}^3$, *let* $\mathbf{T}_p$ *be a unit tangent vector at p on C, and let* **n** *be a vector-valued function defined in a neighborhood of p that gives a continuous choice of unit normal vector. If* **n** *is differentiable along C, then the directional derivative* $\mathbf{T}_p\mathbf{n}$ *is in* T_pM *and the normal curvature* $\kappa_\mathbf{n}$ *of C at the point p is given by*

$$\boldsymbol{\kappa}_\mathbf{n} = \langle \mathbf{T}_p, -\mathbf{T}_p\mathbf{n} \rangle \, \mathbf{n}.$$

[Hint: Use parametrization by arclength and note that $\boldsymbol{\kappa}_\mathbf{n} = \langle \boldsymbol{\kappa}, \mathbf{n} \rangle \, \mathbf{n}$. Let $\gamma(s)$ be a parametrization of the curve C by arclength such that $\gamma(0) = p$. Then the curvature of the curve at p is, by definition,

$$\boldsymbol{\kappa} = \frac{d}{ds}\mathbf{T}(\gamma(s))_{s=0}.$$

And

$$\mathbf{T}_p\mathbf{n} = \frac{d}{ds}\mathbf{n}(\gamma(s))_{s=0}.$$

Also, note that along the curve:

$$\langle \mathbf{T}(\gamma(s)), \mathbf{n}(\gamma(s)) \rangle = 0.$$

Be sure you see why these statements are true and then differentiate the last expression with respect to s.]

b. *Find geometric meaning in the expression* $\langle \mathbf{T}_p, -\mathbf{T}_p\mathbf{n} \rangle$ *by relating it to Problems* **2.3.d** *and* **4.1**.

Here we see a first hint of why it may be possible for the normal curvature (the curvature due to the curving of the surface) to produce an intrinsic quantity because, even though $\mathbf{n}$ is an extrinsic quantity, its derivative $\mathbf{T}_p\mathbf{n}$ (being the derivative of a unit vector) is a tangent vector at p (*Why?*) and thus is intrinsic.

***c.** *Find a simple smooth surface on which, at some point p, the normal* **n** *is not differentiable in some directions.*

d. *Use Part* **a** *to show that on a sphere of radius R, the (scalar) normal curvature of every curve at every point is 1/R.*

Since $\boldsymbol{\kappa}_\mathbf{n}$ depends only on the unit tangent vector $\mathbf{T}_p$ we see that the normal curvature is the same for all curves through p that have the same unit tangent vector (that is, that go in the same direction). Thus we can speak of the normal curvature in the direction $\mathbf{T}$ and write

$$|\boldsymbol{\kappa}_\mathbf{n}(\mathbf{T})| = \kappa_n(\mathbf{T}) = \langle \mathbf{T}, -\mathbf{T}\mathbf{n} \rangle.$$

Let us illustrate by applying this to the strake. We can calculate the normal to strake at the point p as

$$\mathbf{n}(\theta, r) = \frac{\mathbf{x}_1 \times \mathbf{x}_2}{|\mathbf{x}_1 \times \mathbf{x}_2|} = \frac{(-k\sin\theta, k\cos\theta, -r)}{\sqrt{k^2 + r^2}}.$$

There is, of course, another unit normal in the opposite direction, but it is conventional to use the above as the choice of normal when there is a local coordinate system.

Now we can differentiate the normal in the coordinate directions:

$$\mathbf{x}_1\mathbf{n} = \frac{d}{dt}\mathbf{n}(\gamma(t))_{t=\theta} = \frac{\partial}{\partial t}\mathbf{n}(t, r)_{t=\theta} =$$

$$= \frac{(-k\cos\theta, -k\sin\theta, 0)}{\sqrt{r^2 + k^2}} = \frac{-k}{\sqrt{r^2 + k^2}}\mathbf{x}_2,$$

and

$$\mathbf{x}_2\mathbf{n} = \frac{d}{dt}\mathbf{n}(\lambda(t))_{t=r} = \frac{\partial}{\partial t}\mathbf{n}(\theta, t)_{t=r} =$$

$$= \frac{((-r)(-k\sin\theta), (-r)(k\cos\theta), -k^2)}{(r^2 + k^2)^{3/2}} = \frac{-k}{(r^2 + k^2)^{3/2}}\mathbf{x}_1$$

Now from Problem **4.7.a** we have that the normal curvatures of the strake in the directions of $\mathbf{x}_1$ and $\mathbf{x}_2$ are:

$$\kappa_{\mathbf{n}}(\mathbf{x}_1) = \langle \mathbf{x}_1, -\mathbf{x}_1\mathbf{n} \rangle = \left\langle \mathbf{x}_1, \frac{k}{\sqrt{r^2 + k^2}}\mathbf{x}_2 \right\rangle = 0,$$

$$\kappa_{\mathbf{n}}(\mathbf{x}_2) = \langle \mathbf{x}_2, -\mathbf{x}_2\mathbf{n} \rangle = \left\langle \mathbf{x}_2, \frac{k}{(r^2 + k^2)^{3/2}}\mathbf{x}_1 \right\rangle = 0.$$

The reader can also find a geometric argument for why these normal curvatures are zero. (See Problem **6.4.a.**) As we shall see in Chapter 6, the fact that both of these normal curvatures are zero does not imply that all normal curvatures are zero. In particular, $\kappa_{\mathbf{n}}(\mathbf{T})$ is not a linear function; in fact, the reader can check (using the bilinearity of the Riemannian metric) that:

$$\kappa_{\mathbf{n}}(a\mathbf{T}) = \langle a\mathbf{T}, -(a\mathbf{T})\mathbf{n} \rangle = a^2\kappa_{\mathbf{n}}(\mathbf{T})$$

and

$$\kappa_n(\mathbf{T}+\mathbf{V}) = \langle \mathbf{T}+\mathbf{V}, -(\mathbf{T}+\mathbf{V})\mathbf{n}\rangle =$$
$$= \kappa_n(\mathbf{T}) + \kappa_n(\mathbf{V}) + \langle \mathbf{T}, -\mathbf{V}\mathbf{n}\rangle + \langle \mathbf{V}, -\mathbf{T}\mathbf{n}\rangle.$$

This will motivate later the definition of the bilinear ***second fundamental form*** as:

$$\text{II}(\mathbf{X},\mathbf{Y}) = \langle \mathbf{X}, -\mathbf{Y}\mathbf{n}\rangle.$$

Geodesic Local Coordinates

If you wish *to find intrinsically a local coordinate chart at the point* p on a smooth surface M in $\mathbb{R}^n$, then you may construct ***geodesic polar coordinates***, $\mathbf{p}(\theta,r)$, or ***geodesic (rectangular) coordinates***, $\mathbf{c}(x,y)$, as follows. (Refer to Figure 4.9.) Choose a base geodesic, γ, with tangent vector $\mathbf{T}$ at p. Then define $\mathbf{p}(\theta,r)$ to be the point that is a distance r from p along the geodesic, which is in the direction making an angle θ (measured counterclockwise from the point of view of the normal to the surface) with $\mathbf{T}$. Assign $\mathbf{c}(x,y)$ to be the point that is obtained by going a distance x along γ in the direction of $\mathbf{T}$ and then going a distance y along the geodesic that is perpendicular to γ at $\mathbf{c}(x,0)$, turning left if y is positive and right if y is negative.

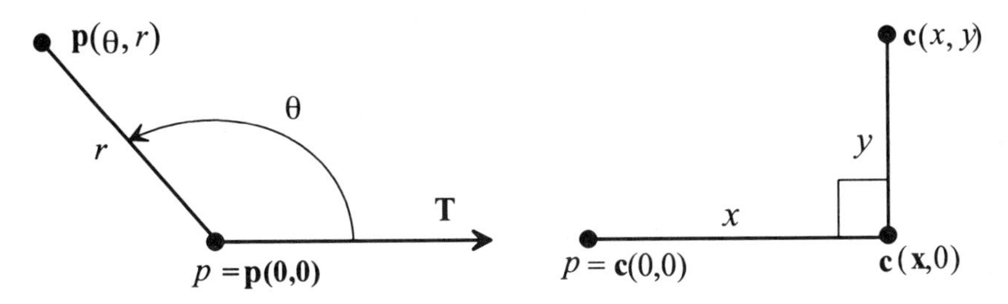

geodesic polar coordinates **geodesic rectangular coordinates**

Figure 4.9.

In geodesic rectangular coordinates, the base curve and the geodesics perpendicular to it are parametrized by arclength: Thus, for every a,b, basis vectors $\mathbf{c}_2(a,b)$ and $\mathbf{c}_1(a,0)$ are unit vectors. However, the coordinate curves $\mathbf{c}(t,b)$, for fixed $b \neq 0$, are in general not parametrized by arclength and $\mathbf{c}_1(a,b) \neq 0$. On the earth the standard north-south-east-west coordinates have the equator (the only latitude circle which is a geodesic) as the base curve. This coordinate system can be changed to be geodesic rectangular coordinates by replacing the angle coordinates with arclength coordinates. The resulting geodesic rectangular coordinate patch will cover all the earth except for the longitude 180°W (= 180°E). A geodesic polar coordinate patch will cover the whole sphere except for one point (the antipodal point to the origin).

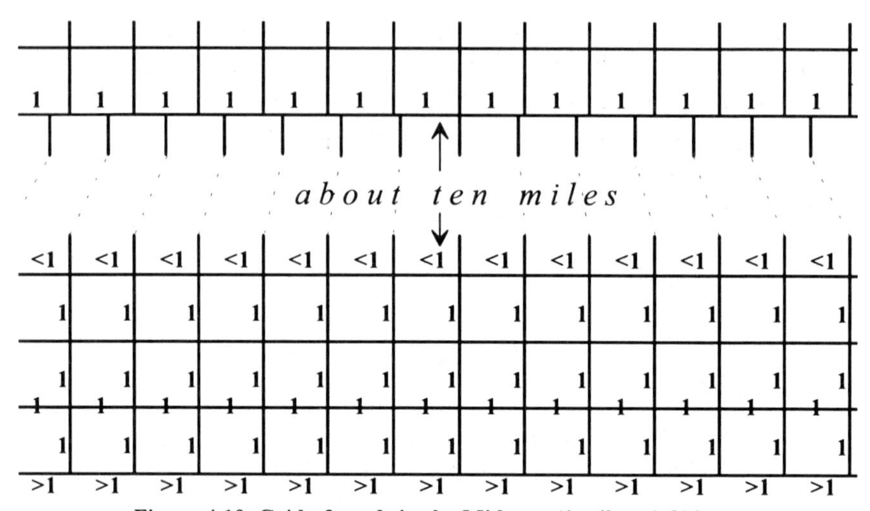

Figure 4.10. Grid of roads in the Midwest (1 mile = 1.61 km).

You can see concrete examples of this phenomenon in many places in the Midwest region of the USA. In the Midwest, typically one road is laid out east-west (as a base curve) and then from this road, at one mile (= 1.61 km) intervals, north-south roads are constructed, and then additional east-west roads at one-mile intervals along these north-south roads. (See Figure 4.10 or maps showing the county roads in some area of the Midwest.) However, as one travels north from the base east-west road, the distance between successive north-south roads becomes, more and more, less than one mile. (If you travel south the opposite happens.) Thus, every ten miles or so, one must make a correction as indicated in Figure 4.10.

PROBLEM 4.8. Differential Operator

We call $\mathbf{x}$ a **C^2 *coordinate patch*** (or **C^2 *local coordinates*) *for*** M if $\mathbf{x}$ is a C^1 coordinate patch and, as a function from $\mathbb{R}^2$ to $\mathbb{R}^n$, $\mathbf{x}$ is C^2 (that is,

x is twice differentiable, and the second partial derivatives are continuous).

You can now prove:

Let $\mathbf{x}(u^1,u^2)$ *be a* C^2 *coordinate patch for the smooth surface* M*. Let* $\mathbf{F}$ *be a real-valued or vector-valued function defined on a neighborhood of* $p = \mathbf{x}(a,b)$ *in* M *such that and* $\mathbf{F} \circ \mathbf{x}$ *is* C^1*. Given tangent vectors* $\mathbf{X}_p$ *and* $\mathbf{Y}_p$ *in* T_pM*, express these tangent vectors in terms of the coordinates and show that*:

a. $\mathbf{X}_p\mathbf{F}$ *does not depend on the choice of curve* $\gamma(t)$ *such that* $\gamma(0) = p$ *and* $\gamma'(0) = \mathbf{X}_p$.

b. *For any scalar a,*

$$(\mathbf{X}_p + \mathbf{Y}_p)\mathbf{F} = \mathbf{X}_p\mathbf{F} + \mathbf{Y}_p\mathbf{F} \text{ and } (a\mathbf{X}_p)\mathbf{F} = a(\mathbf{X}_p\mathbf{F}).$$

[Note: If α, β, γ are curves on the surface such that

$$\alpha'(0) = \mathbf{X}_p,\ \beta'(0) = \mathbf{Y}_p,\ \gamma'(0) = \mathbf{X}_p + \mathbf{Y}_p,$$

and

$$\alpha(0) = \beta(0) = \gamma(0) = p,$$

then in $\mathbb{R}^n$ it is possible to specify that $\gamma(s) = \alpha(s) + \beta(s)$, but this is NOT possible in general on a surface. *Be sure you see why*.]

c. *If* f *is a* C^1 *real-valued function, then*

$$\mathbf{X}_p(f\mathbf{F}) = (\mathbf{X}_p f)\mathbf{F} + f(\mathbf{X}_p\mathbf{F}).$$

[Hint: this is the product rule.]

d.

$$\mathbf{x}_1(a,b)\mathbf{x}_2 = \mathbf{x}_2(a,b)\mathbf{x}_1.$$

Each part of this problem is almost just a matter of notation. If you rewrite the equations in terms of the local coordinates, then you will see that they hold. This is more a notational problem and not a technical problem. Note that $\mathbf{x}_j$ is a (tangent) vector-valued function of the coordinates (u^1,u^2) or, equivalently, is a function of the points $q = \mathbf{x}(u^1,u^2)$ in M.

Thus $\mathbf{x}_1(a,b)$ can also be written $\mathbf{x}_1(\mathbf{x}(a,b))$ as a tangent vector at the point $p = \mathbf{x}(a,b)$ and

$$\mathbf{x}_1(a,b)\mathbf{x}_2 = \lim_{h\to 0}\frac{\mathbf{x}_2(\mathbf{x}(a+h,b)) - \mathbf{x}_2(\mathbf{x}(a,b))}{h} =$$

$$= \lim_{h\to 0}\frac{\mathbf{x}_2(a+h,b) - \mathbf{x}_2(a,b)}{h} = \frac{\partial}{\partial u^1}\mathbf{x}_2(u^1,b)_{u^1=a}$$

is the directional derivative of the function $\mathbf{x}_2$ with respect to $\mathbf{x}_1(a,b)$. Thus,

$$\mathbf{x}_1(a,b)\mathbf{x}_2 = \frac{\partial}{\partial u^1}\mathbf{x}_2(u^1,b)_{u^1=a} = \frac{\partial}{\partial u^1}\left(\frac{\partial}{\partial u^2}\mathbf{x}(u^1,u^2)\right)_{(u^1,u^2)=(a,b)}.$$

It is usual to define:

$$\mathbf{x}_{ij} = \mathbf{x}_i\mathbf{x}_j = \frac{\partial}{\partial u^i}\frac{\partial}{\partial u^j}\mathbf{x}.$$

e. *Calculate* $\mathbf{x}_{12}$ *and* $\mathbf{x}_{21}$ *for the standard coordinate system on a sphere and on a strake.*

On the sphere of radius r we calculate that

$$\mathbf{x}_{12} = \mathbf{x}_{21} = (-r\sin\theta\cos\phi,\ r\cos\theta\cos\phi,\ 0) = (\cot\phi)\mathbf{x}_1.$$

The length of the tangent vector in the latitudinal (east-west) direction of $\mathbf{x}_1$ starts off at r^2 on the equator but decreases as you move toward either pole; and $\mathbf{x}_{21}$ is the rate of change of $\mathbf{x}_1$ as you move southward along a longitude. Also, note that $\mathbf{x}_{12}$ is the rate of change of the tangent vector in the longitudinal direction of $\mathbf{x}_2$ as you move westward along a latitude circle and, even though the length of $\mathbf{x}_2$ is constantly r^2, its direction is changing. I urge the reader to investigate this phenomenon on the sphere until it becomes as natural and comfortable as possible.

Problem 4.9. Metric in Geodesic Coordinates

Explain each step in the following argument.

Let $\mathbf{x}(u^1,u^2)$ be geodesic rectangular coordinates, $\mathbf{c}(x,y)$, or geodesic polar coordinates, $\mathbf{p}(\theta,r)$, as in Figure 4.9 above. According to Problem **4.3** the Riemannian metric can be expressed in local coordinates as the matrix:

$$g = (g_{ij}) = \begin{pmatrix} \langle \mathbf{x}_1, \mathbf{x}_1 \rangle & \langle \mathbf{x}_1, \mathbf{x}_2 \rangle \\ \langle \mathbf{x}_2, \mathbf{x}_1 \rangle & \langle \mathbf{x}_2, \mathbf{x}_2 \rangle \end{pmatrix}.$$

a. From the definition of geodesic coordinates, for constant a, the geodesic curves $\mathbf{x}(a, u^2)$ are parametrized by arclength and thus

$$g_{22}(u^1, u^2) = 1. \ (\textit{Why?})$$

b. We now need to find $g_{12}(u^1, u^2)$. To do this we first differentiate:

$$\frac{\partial}{\partial u^2} g_{12}(u^1, u^2) = \mathbf{x}_2 \langle \mathbf{x}_1(u^1, u^2), \mathbf{x}_2(u^1, u^2) \rangle =$$
$$= \langle \mathbf{x}_{21}, \mathbf{x}_2 \rangle + \langle \mathbf{x}_1, \mathbf{x}_{22} \rangle. \ (\textit{Why?})$$

Now, since $\mathbf{x}_2$ is a unit vector

$$\langle \mathbf{x}_{21}, \mathbf{x}_2 \rangle = \langle \mathbf{x}_{12}, \mathbf{x}_2 \rangle = \left\langle \frac{\partial}{\partial u^1} \mathbf{x}_2, \mathbf{x}_2 \right\rangle = 0. \ (\textit{Why?})$$

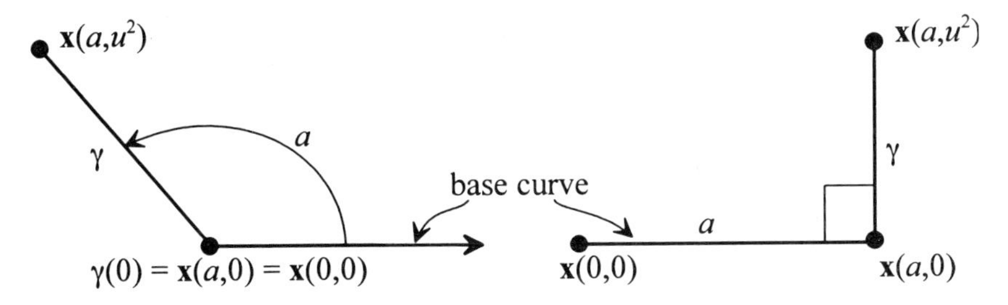

Figure 4.11. Second coordinate curves.

c. Now we focus on the second coordinate curve

$$\gamma(u^2) = \mathbf{x}(a, u^2),$$

(see Figure 4.11). Since $\gamma(u^2) = \mathbf{x}(a, u^2)$ is parametrized by arclength, its unit tangent vector is $\mathbf{x}_2(a, u^2)$ and thus

$$\mathbf{x}_{22} = \frac{\partial^2}{(\partial u^2)^2} \mathbf{x}(a, u^2) = \frac{\partial}{\partial u^2} \mathbf{x}_2(a, u^2) = \kappa$$

is its (extrinsic) curvature vector. (*Why?*) Since the curve is a geodesic, its curvature vector must be parallel to the normal to the surface. Thus

$$\langle \mathbf{x}_1, \mathbf{x}_{22} \rangle = 0 \text{ and therefore } \frac{\partial}{\partial u^2} g_{12}(a, u^2) = 0.$$

We can then conclude that $g_{12}(a,u^2)$ is a constant independent of u^2. (*Why?*)

d. By definition of geodesic rectangular coordinates,

$$g_{12}(u^1,0) = \langle \mathbf{x}_1(u^1,0), \mathbf{x}_2(u^1,0) \rangle = 0. \ (\textit{Why?})$$

For geodesic polar coordinates, $\mathbf{x}(u^1,0) = \mathbf{p}(\theta,0) = \mathbf{p}(0,0)$, a constant. Thus, again,

$$g_{12}(u^1,0) = \langle \mathbf{x}_1(u^1,0), \mathbf{x}_2(u^1,0) \rangle = 0. \ (\textit{Why?})$$

e. We can now conclude that

$$g_{12}(u^1,u^2) = 0, \text{ for all } u^1 \text{ and } u^2. \ (\textit{Why?})$$

Thus, for geodesic rectangular or polar coordinates:

$$g(u^1, u^2) = \begin{pmatrix} (h(u^1, u^2))^2 & 0 \\ 0 & 1 \end{pmatrix},$$

where $h(u^1,u^2) = |\mathbf{x}_1(u^1,u^2)| > 0$.

We will use this representation of the Riemannian metric to find explicit intrinsic calculations of the Gaussian curvature in local coordinates in Chapter 7.

Chapter 5

Area, Parallel Transport and Intrinsic Curvature†

In Chapters 3 and 4 we were developing extrinsic descriptions of the intrinsic curvature of a curve on a surface. In this chapter we will develop intrinsic descriptions of both the ***intrinsic curvature of a curve*** on a surface and the ***intrinsic curvature of a surface***. This intrinsic curvature will be developed by investigating the relationships between surface area, normal curvature, and ***parallel transport***, a notion of local parallelism that is definable on all surfaces. We first develop these connections on a sphere that has known curvature and then use the results on the sphere to motivate the discussion on general surfaces. As an important part of this development, we introduce the notion of ***holonomy***, which also has many other applications in modern differential geometry and engineering.

PROBLEM 5.1. The Area of a Triangle on a Sphere

DEFINITION: On any surface we will call a triangle a ***geodesic triangle*** if its three sides are geodesic segments.

a. *Let* Δ *be a geodesic triangle on a sphere. Show that the formula*

$$\text{Area}\,(\Delta) = A/4\pi[\,\textstyle\sum \beta_i - \pi\,] = A/4\pi[\,2\pi - \textstyle\sum \alpha_i\,]$$

holds, where A is the area of the sphere, β_i *are the interior angles, and* α_i *are the exterior angles of the triangle in radians.* The quantity $\Sigma\beta_i - \pi$ is called the ***excess*** of Δ.

† The first part of this chapter is taken (somewhat revised) from the author's *Experiencing Geometry on Plane and Sphere* [**Tx**: Henderson]. It is used here with the permission of Prentice-Hall, Inc.

We offer the following hint as a way to approach this problem: Find the area of a biangle (lune) with angle θ. (A ***biangle*** or ***lune*** is one of the connected regions between two great circles.) Notice that the great circles that contain the sides of the triangle divide the sphere into overlapping biangles. Focus on the biangles determined by either the interior angles or the exterior angles.

This is one of the problems that you almost certainly must do on an actual sphere. There are too many things to see, and the drawings we make on paper distort lines and angles too much. The best way to start is to make a small triangle on a sphere, and extend the lines of the triangle to complete great circles. Then look at the results. You will find an identical triangle on the other side of the sphere, and you can see several lunes extending out from the triangles. The key to this problem is to put everything in terms of areas that you know. *Find the areas of the lunes*, as noted above. After that, it is simply a matter of adding everything up properly.

We know that the area of the whole sphere is $4\pi R^2$, where R is the (extrinsic) radius of the sphere. With this additional information we can rewrite the formula of Problem **5.1.a**:

$$\text{Area}\ (\Delta) = [\ \sum \beta_i - \pi\ ]\ R^2 = [\ 2\pi - \sum \alpha_i\ ]\ R^2.$$

b. *As R^2 goes to infinity, the sphere locally becomes closer and closer to a plane. Look at a very small geodesic triangle on a sphere with a very large radius. For this triangle, compare the formula from* **5.1.a** *with what you know about plane triangles.*

c. *Conclude that a sphere is not locally isometric with a plane.*

Introducing Parallel Transport

Imagine that you are walking along a straight line or geodesic, carrying a horizontal rod that makes a fixed angle with the line you are walking on. If you walk along the line maintaining the direction of the rod relative to the line constant, then you are performing a ***parallel transport*** of that "direction" along the path. (See Figure 5.1.)

To express the parallel transport idea, it is common terminology to say that:

- ***r′*** is a parallel transport of ***r*** along ***l***;
- ***r*** is a parallel transport of ***r′*** along ***l***;
- ***r*** and ***r′*** are parallel transports along ***l***;
- ***r*** can be parallel transported along ***l*** to ***r′***; or,
- ***r′*** can be parallel transported along ***l*** to ***r***.

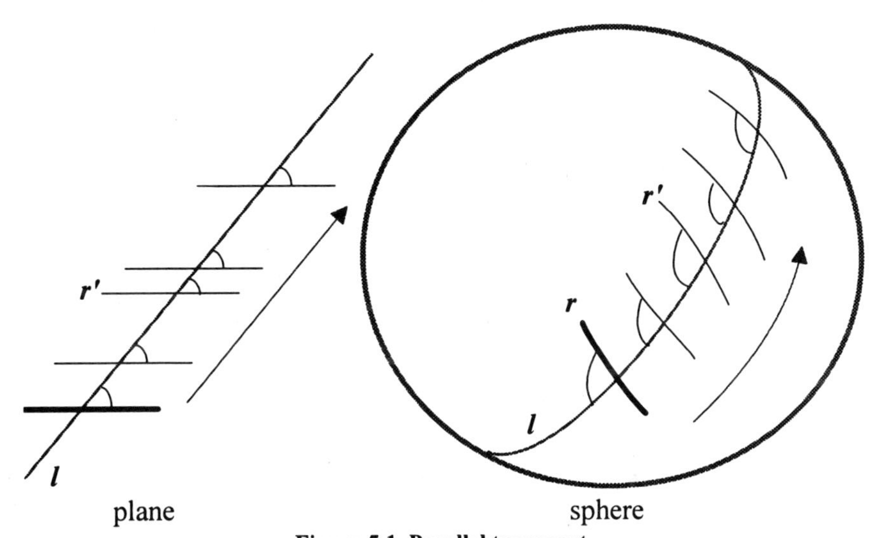

Figure 5.1. Parallel transport.

Parallel transport has become an important notion in Differential Geometry, Physics, and Mechanics. One important aspect of Differential Geometry is the study of properties of spaces (surfaces) from an intrinsic point of view. As we have seen, in general, it is not possible to have a global notion of direction from which we are able to determine when a direction (or vector) at one point is the same as a direction (or vector) at another point. However, we can say that they have the same direction *with respect to* a geodesic *g* if they are parallel transports of each other along *g*. Parallel transport can be extended to arbitrary curves as we shall discuss in Problem **5.4**. With this notion it is possible to talk about how a particular vector quantity changes intrinsically along a curve (covariant differentiation). In general, covariant differentiation is useful in the areas of physics and mechanics. In physics, the notion of parallel transport is central to some of the theories that have been put forward as possible candidates for a "Unified Field Theory," a hoped for, but as yet unrealized, theory that would unify all known physical laws about the forces of nature.

The Holonomy of a Small Geodesic Triangle

Let us imagine that we parallel transport a vector along a piecewise geodesic curve (such as a geodesic triangle) in a surface M. Since we do not want the parallel transported vector to change direction intrinsically as we parallel transport along the curve, we must change the angle between the transported vector and the velocity vector in order to undo the effect of the change in the direction of the curve. Thus the angle between the transported vector and the velocity vector changes at every vertex by the amount of the exterior angle. (See Figure 5.2.)

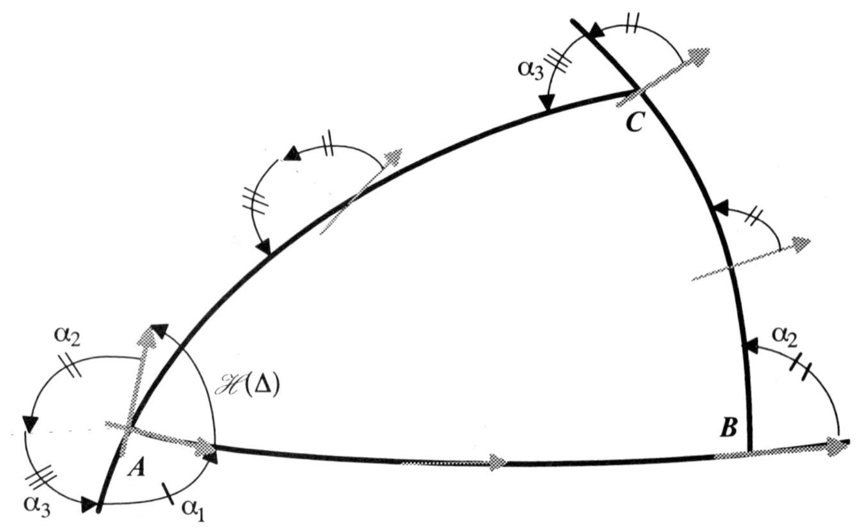

Figure 5.2. Holonomy of a geodesic triangle.

DEFINITION: The ***holonomy***† ***of a small*** (that is, contained in an open hemisphere) ***geodesic triangle***, $\mathcal{H}(\Delta)$, is defined as follows:

If you parallel transport a vector (a directed geodesic segment) counterclockwise around the three sides of a small

†Some books define holonomy as [(our holonomy) – 2π]. I chose not to use this definition, because it would result in the holonomy of a planar triangle being egual to -2π. I think it is more natural for the holonomy of a planar triangle to be zero.

triangle, then the ***holonomy*** *of the triangle is the smallest angle measured counterclockwise from the original position of the vector to its final position.* (See Figure 5.2.)

Later we will show how to extend this definition to apply to all triangles and polygons. By studying the above picture the reader should be able to see that:

The holonomy of a small geodesic triangle is equal to 2π *minus the sum of the exterior angles.*

Let β_1, β_2 and β_3 be the interior angles of the triangle and α_1, α_2, α_3, the exterior angles, then algebraically the statement above can be written as:

$$\mathscr{H}(\Delta) = 2\pi - (\alpha_1 + \alpha_2 + \alpha_3) = (\beta_1 + \beta_2 + \beta_3) - \pi.$$

The quantity $(\beta_1 + \beta_2 + \beta_3) - \pi$ is traditionally called the ***excess*** of the triangle.

Note one consequence of this formula :

The holonomy does not depend on either the vertex or the vector we start with.

We can write the result from Problem **5.1.a** in the following form:

For a small geodesic triangle on the sphere:

$$\mathscr{H}(\Delta) = 2\pi - (\alpha_1 + \alpha_2 + \alpha_3) = A(\Delta)\, 4\pi/A = A(\Delta)\, R^{-2}.$$

The formula

$$2\pi - (\alpha_1 + \alpha_2 + \alpha_3) = A(\Delta)\, R^{-2}$$

is called the ***Gauss-Bonnet Formula*** (for triangles). The quantity R^{-2} is traditionally called the ***Gaussian curvature***, or ***intrinsic curvature***, or just plain ***curvature*** of the sphere. Can you see how this result gives the bug an intrinsic way of determining the extrinsic quantity R and the curvature

$$R^{-2} = \mathscr{H}(\Delta)/A(\Delta)\ ?$$

As soon as we have defined the notion of holonomy on a surface then we can define the intrinsic curvature, $K(p)$, at a point on the surface to be

$$K(p) = \lim_{n\to\infty} \mathscr{H}(R_n)\, /\, A(R_n),$$

where $\{R_n\}$ is a sequence of small (geodesic) triangles that converge to p. (See Problem **5.6**.) In Problem **6.4** we show that this intrinsic curvature is the same on C^2 surfaces as the usually defined Gaussian curvature.

The Gauss-Bonnet Formula not only holds on the sphere for small triangles but can be extended to any *small* (contained in an open hemisphere) *simple* (nonintersecting) *polygon* (a closed curve consisting of a finite number of geodesic segments) on a sphere. We will need the result for small simple polygons on the sphere in order to make the transition from the sphere to arbitrary surfaces.

We shall prove the Gauss-Bonnet Formula for polygons on the sphere by first dividing the polygon into triangles as in the following:

Problem 5.2. Dissection of Polygons into Triangles

Prove that every simple polygon on a sphere and plane can be dissected into small triangles without adding extra vertices.

Consider this problem on both the plane and sphere. The difficulty in this problem lies in divising a method that works for all simple polygons, including very general or complex ones, like in Figure 5.3.

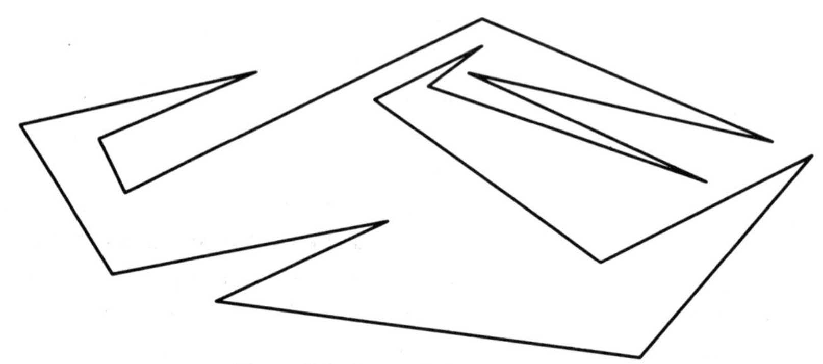

Figure 5.3. General simple polygon.

You may be tempted to try to connect nearby vertices to create triangles, but how do we know that this is always possible? How do you know that in any polygon there is even one pair of vertices that can be joined in the interior? The polygon may be so complex that parts of it get in the way of what you are trying to connect. You might start by giving a convincing argument that there is at least one pair of vertices that can be joined by a segment in the interior of the polygon. In order to see that there is something to prove here, look at Figure 5.4, which shows a polyhedron in 3-space with **no** pair of vertices that can be joined in the interior. The polyhedron consists of eight triangular faces and six vertices. Each vertex is joined by an edge to four of the other vertices and the straight line segment joining it to the fifth vertex lies in the exterior of the polygon. Therefore it is impossible to dissect this polyhedron into tetrahedra without adding extra vertices.

This example and some history of the problem are discussed in [**P**: Eves], page 211. Use Computer Exercise 5.2 to display images of this polyhedron, which may then be viewed from various perspectives.

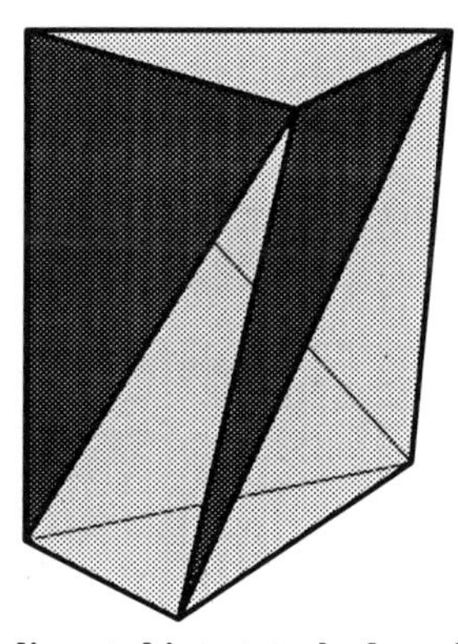

Figure 5.4. Cannot be dissected into tetrahedra without adding vertices.

Note that there is at least one convex vertex (a vertex with interior angle less than π) on every polygon (in fact it is not too hard to see that there must be at least three such vertices). For this and other problems, pick any (straight) line in the exterior of the polygon and parallel transport it towards the polygon until it first touches the polygon. It is easy to see that the line must now be intersecting the line at a convex vertex.

PROBLEM 5.3. Gauss-Bonnet for Polygons on a Sphere

DEFINITION: The holonomy of a small simple (geodesic) polygon, $\mathscr{H}(\Gamma)$, is defined as follows:

If you parallel transport a vector (a directed geodesic segment) counterclockwise around the sides of a small simple polygon, then the holonomy of the polygon is the smallest angle measured counterclockwise from the original position of the vector to its final position.

If you walk around a polygon with the interior of the polygon on the left, the exterior angle at a vertex is the change in the direction at that vertex. This change is positive if you turn counterclockwise and negative if you turn clockwise. (See Figure 5.5.)

Show that for Γ*, a small simple polygon on a sphere,*

$$\mathscr{H}(\Gamma) = A(\Gamma)\, 4\pi/A = A(\Gamma)\, R^{-2} = 2\pi - \Sigma\, \alpha_i\,,$$

where $\Sigma\alpha_i$ *is the sum of the exterior angles of the polygon.*

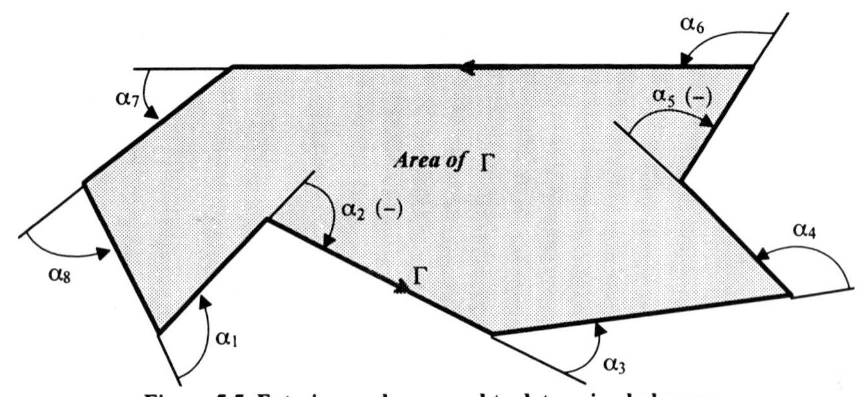

Figure 5.5. Exterior angles as used to determine holonomy.

Outline of a proof: Divide the polygon into small triangles. It is possible to do so by constructing geodesic segments in the interior of the polygon without adding any new vertices (see Problem **5.2**). Do this problem in two steps, and in each of these steps use the definition of holonomy in terms of parallel transport: First, by removing the small triangles one at a time, show that the holonomy of the polygon is the sum of the holonomies of the triangles. Second, check directly that

$$\mathscr{H}(\Gamma) = 2\pi - \Sigma\, \alpha_i\,.$$

PROBLEM 5.4. Parallel Fields and Intrinsic Curvature

When we parallel transport a vector along a piecewise geodesic curve γ in a surface M, then the angle between the transported vector and the velocity vector changes at every vertex by the amount of the exterior angle. (See Figure 5.6.)

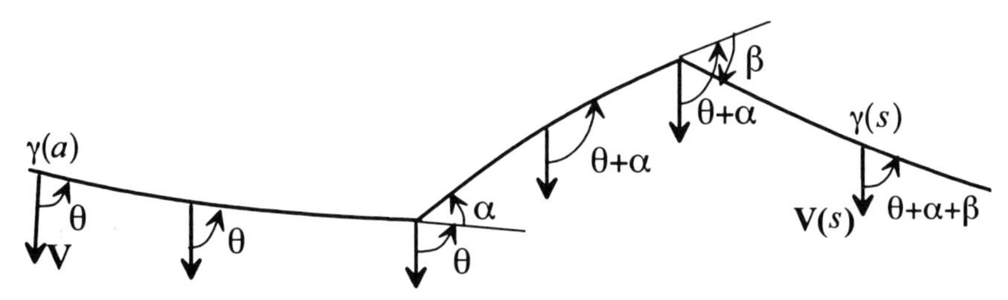

Figure 5.6. Parallel vector field along a piecewise geodesic curve.

Since we do not want the parallel transported vector to change direction as we parallel transport along the curve, we must change the angle between the transported vector and the velocity vector in order to undo the effect of the change in the direction of the curve. If **V** is a vector tangent to the surface at $\gamma(a)$ such that θ is the angle from **V** to the velocity vector $\gamma\,'(a)$, then we can define

$$\mathbf{V}(s) \equiv P(\gamma,\gamma(a),\gamma(s))\mathbf{V},$$

the ***parallel transport of* V *along* γ *to the point*** $\gamma(s)$, to be the vector (with the same length as **V**) tangent to the surface at $\gamma(s)$ such that the angle from $\mathbf{V}(s)$ to $\gamma\,'(s)$ is equal to θ plus the sum of the exterior angles of γ between $\gamma(a)$ and $\gamma(s)$. (See Figure 5.6.) If we let s vary over all of γ, then the vectors, $\mathbf{V}(s) = P(\gamma,\gamma(a),\gamma(s))\mathbf{V}$, are called a ***parallel vector field*** along γ. We say that $\mathbf{W}(s)$ is simply a ***vector field*** along γ, if $\mathbf{W}(s)$ is a tangent vector in the tangent space at $\gamma(s)$ for each s.

a. *Show that the following statements are equivalent for a vector field* $\mathbf{V}(s)$ *on a piecewise geodesic curve* γ *parametrized by arclength:*

i. $\mathbf{V}(s)$ *is a parallel vector field along g as defined above.*

ii. *The derivative* $\frac{d}{ds}\mathbf{V}(s)$ *is perpendicular to the tangent plane at* $\gamma(s)$.

[Hint: Note that the derivative $\frac{d}{ds}\mathbf{V}(s)$ is the same as the directional derivative $(\gamma\,'(s))\mathbf{V}$. Use the fact that on a geodesic the angle between the transported vector and the velocity vector is constant. At the corners of a piecewise geodesic curve use one-sided derivatives:

$$\frac{d}{ds+}\mathbf{V}(s) = \lim_{h\to 0^+}\frac{1}{h}[\mathbf{V}(s+h) - \mathbf{V}(s)]$$

and

$$\frac{d}{ds-}\mathbf{V}(s) = \lim_{h\to 0^+}\frac{1}{-h}[\mathbf{V}(s+h) - \mathbf{V}(s)],$$

where h is always positive.]

This motivates an ***extrinsic definition of parallel transport*** for arbitrary curves:

> **Definition**: $\mathbf{V}(s)$ is a ***parallel vector field*** along the curve γ in the surface M if at each point $\gamma(s)$ the derivative $\frac{d}{ds}\mathbf{V}(s)$ is perpendicular to the tangent space $T_{\gamma(s)}M$.

In Chapter 2 we showed that if $\gamma(s)$ is a curve (parametrized by arclength) on the plane, then the magnitude of the curvature $|\kappa(s)|$ is equal to the magnitude of the rate of change of the angle between $\gamma\,'(s)$ and the horizontal. This is possible on the plane because the horizontal gives a global notion of parallel, and this is precisely what is impossible on more general surfaces. But if we have a parallel vector field $\mathbf{V}(s)$ along γ on a surface, then we might expect that the magnitude of the intrinsic curvature is equal to the rate of change of the angle between $\mathbf{V}(s)$ and $\gamma\,'(s)$. In fact we can prove:

b. *If $\mathbf{V}(s)$ is a smooth vector field along the smooth C^2 curve γ on the surface M, then the following statements are equivalent:*

i. *The derivative $\frac{d}{ds}\mathbf{V}(s)$ is perpendicular to the tangent plane at $\gamma(s)$.*

ii. *If $\theta(s)$ is the counterclockwise angle from $\mathbf{V}(s)$ to the (unit) tangent vector $\gamma\,'(s)$, then*

$$\frac{d}{ds}\theta(s) = \kappa_g(s),$$

where $\kappa_g(s) = \pm|\boldsymbol{\kappa}_g(s)|$ *is the (scalar) geodesic curvature of* γ, *positive if* γ *is turning counterclockwise at* $\gamma(s)$. (See Figure 5.7.)

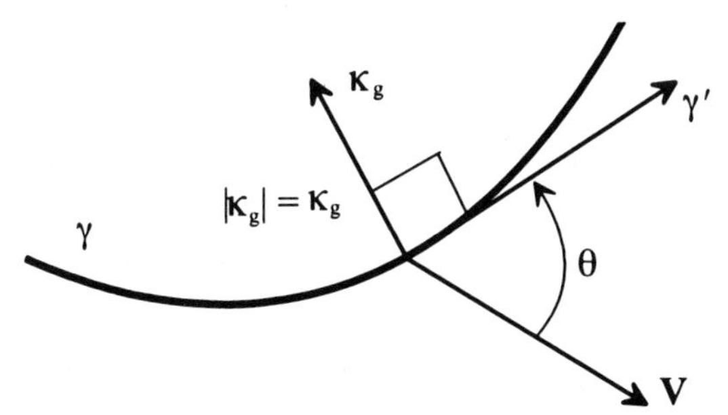

Figure 5.7. Positive geodesic curvature.

[Hint: Differentiate

$$\langle \mathbf{V}(s), \gamma'(s) \rangle = \cos\theta$$

and note that

$$\gamma''(s) = \boldsymbol{\kappa}(s) = \boldsymbol{\kappa}_g(s) + \boldsymbol{\kappa}_n(s).]$$

c. *Let* γ *be a piecewise smooth curve with finite length which is parametrized by arclength. Let* γ_i *be a sequence (which converges pointwise to* γ*) of piecewise geodesics* $\{\gamma_i\}$ *whose vertices lie on* γ. *Show that the parallel transport of* $\mathbf{V}$ *along* γ *to* $p = \gamma(s)$ *is the limit of the tangent vectors* $\mathbf{V}_i$ *at p, which are parallel transports along* γ_i. *This gives an* ***intrinsic definition of parallel transport****.* (See Figure 5.8.)

[Hint: Follow these steps:

1. Let $\mathbf{V}(s) \equiv \lim \mathbf{V}_i$ and argue that is perpendicular to the tangent plane $\mathbf{T}_p$. You may need to use the fact that the tangent planes vary continuously.
2. Then use Part **b.ii** to show that there is a unique vector field satisfying this property.]

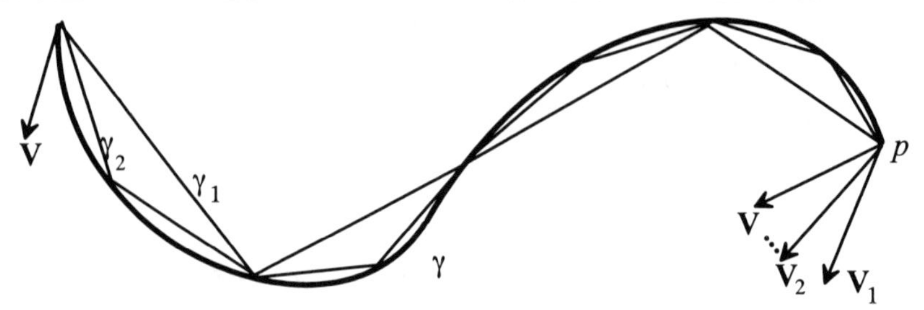

Figure 5.8. Defining parallel transport on a smooth curve.

Now we can now give an ***intrinsic definition of intrinsic (geodesic) curvature*** as:

DEFINITION: If **V**(s) is a parallel vector field along the curve γ in the surface M and if, at each point $\gamma(s)$, $\theta(s)$ is the counterclockwise angle from **V**(s) to $\gamma'(s)$, then

$$\kappa_g(s) = \frac{d}{ds}\theta(s).$$

d. *Derive the following corollaries:*

i. *Show that the smooth curve γ is a geodesic if and only if $\gamma'(s)$ is a parallel vector field along γ.*

ii. *For a small piecewise smooth closed curve γ, the holonomy defined in terms of parallel transport is equal to*

$$\mathcal{H}(\gamma) = 2\pi - [\,\Sigma\,\alpha_i + \textstyle\int_\gamma \kappa_g\,].$$

We can now extend the notion of holonomy and the Gauss-Bonnet formula (Problem **5.3**) on the sphere to piecewise smooth curves.

DEFINITION: The ***holonomy of a simple closed piecewise smooth curve*** is defined to be:

$$\mathcal{H}(\gamma) = 2\pi - [\,\Sigma\,\alpha_i + \textstyle\int_\gamma \kappa_g\,].$$

If γ_i is a sequence (which converges pointwise to γ) of piecewise geodesic closed curves $\{\gamma_i\}$ whose vertices lie on γ, then

$$\lim_{i\to\infty} \mathcal{H}(\gamma_i) = \mathcal{H}(\gamma)$$

and thus we conclude:

For any simple closed piecewise smooth curve γ *on the sphere,*

$$\mathcal{H}(\gamma) = 2\pi - [\Sigma\, \alpha_i + \int_\gamma \kappa_g] = A(\gamma)\, R^{-2},$$

where $A(\gamma)$ *is the area on the left hand side of* γ.

PROBLEM 5.5. Holonomy on Surfaces

Find the holonomy of the following regions—make note of which holonomies are positive, which are negative, and which are zero.

a. *Find the holonomy of a geodesic triangle on a cylinder or cone such that the area of the triangle is finite and does not contain the cone point.*

[Hint: Note that you can cut the cone or cylinder and flatten it to a plane.]

b. *Find the holonomy of an intrinsic circle with center at the cone point on a cone with cone angle* α. *Note when the holonomy is positive and when negative.*

c. *Find the holonomy of the two regions marked A and B on the torus in Figure 5.9.*

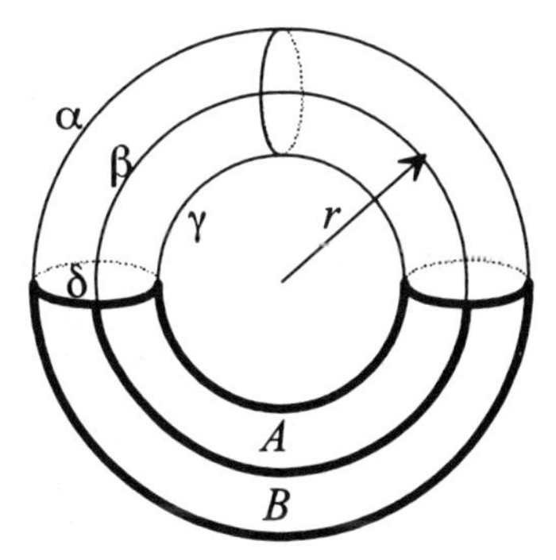

Figure 5.9. Holonomy on a torus.

[Hint: The curves α, δ, and γ are geodesics. (*Why?*) The curve β has its extrinsic curvatures tangent to the surface, and therefore, the extrinsic and intrinsic curvatures coincide.]

d. *Find the holonomy, on a strake, of the region bounded by four coordinate curves:*

$$\mathbf{x}(\theta,t),\ \mathbf{x}(t,r),\ \mathbf{x}(\theta+\Delta,t),\ \mathbf{x}(t,r+\Delta).$$

[Hint: Note that the straight horizontal coordinate curves are geodesics and that the helical coordinate curves have their extrinsic curvatures equal to their intrinsic curvatures. Your answer should be negative.]

We can say that regions on a surface have ***negative, zero, or positive curvature*** depending on whether small polygons in the region have holonomy that is negative, zero, or positive. Computer Exercise 5.5 uses the intrinsic formula for intrinsic (Gaussian) curvature developed in Problem **7.1** to allow you display a surface with different colors marking the regions of positive, negative, and zero curvature.

PROBLEM 5.6. Holonomy Explains Foucault's Pendulum

We now give a physical example of holonomy at work. Jean Foucault, about 1851, built a pendulum consisting of a heavy iron ball on a wire 200 feet long. Foucault observed that, as time passed, the pendulum rotated, eventually returning to its original position after a period of

$$T = \frac{24}{\sin\phi},\quad \phi = \text{latitude},$$

the latitude on which the pendulum sits. Most science museums and physics departments have an example of such a pendulum which has come to be known as ***Foucault's Pendulum*** (or the ***Foucault pendulum***).

Imagine a sphere S with the same axis as the earth, which always stays with the earth but does not rotate on its axis. Because of the rotation of the earth, the pendulum moves around this stationary sphere along a latitude circle. Since the wire is very long compared to the small swing of the ball, we can consider the ball to be moving along a tangent line to the sphere. If we pick consistently one of the two possible directions along this line, then the swing plane of Foucault's Pendulum determines a unit tangent vector to the sphere. Over time this defines a vector field along the latitude circle. Since the pendulum is being moved very slowly around the sphere, the consequent centripetal force on the pendulum is negligible. Thus, the only significant force on the pendulum is due to gravity, which operates in a direction that is in the swing plane of the pendulum and perpendicular to the surface.

For the following problems, it may be helpfulto know that the radius of the earth is approximately 6360 km.

a. *Argue that the vector field determined by the swing plane of Foucault's Pendulum is parallel along the latitude circle. Thus, after 24 hours (one revolution), the swing plane of Foucault's Pendulum rotates by an amount of the holonomy of the latitude circle minus* 2π.

[Hint: Look at Figure 5.2.]

b. *Calculate the holonomy of the latitude circle with angle* ϕ, *measured from the Equator.*

c. *What should be the period of rotation of Foucault's Pendulum?*

d. *What would happen to Foucault's Pendulum at the North Pole? on the Equator?*

e. *Use the above to calculate the area of the Earth's surface above the latitude* ϕ.

PROBLEM 5.7. *Intrinsic Curvature of a Surface*

DEFINITION: ***Intrinsic curvature*** $K(p)$ at a point p on any smooth surface can be defined intrinsically as:

$$K(p) = \lim_{n\to\infty} \mathscr{H}(R_n) \,/\, A(R_n),$$

where $\{R_n\}$ is a sequence of small (geodesic) polygons that converge to p.

It will follow from Problem **6.4** that this definition does not depend on which sequence of polygons is chosen.

a. *What is the intrinsic curvature at a point on a cylinder? What is the intrinsic curvature at points on a cone?*

b. *What is the intrinsic curvature at points on a sphere?*

***c.** *What is the intrinsic curvature at a point* **x**(a,b) *on a strake?*

[Hint: Use Problems **4.5.d** and **5.5.d**. Instead of the closed form expression for the area you may find it easier to leave it (at least partially) in its integral form and then evaluate the limit by using L'Hôpital's Rule.]

d. *What is the intrinsic curvature at a point on the annular hyperbolic plane?*

[Hint: Note that since the surface is constructed the same everywhere (as $\delta \to 0$), it is ***homogeneous*** (that is, intrinsically and geometrically, every point has a neighborhood that is isometric to a neighborhood of any other point). Thus, the intrinsic curvature is constant. Pick a region that crosses the circular edge between two strips and then let $\delta \to 0$. (See Figure 5.10.) Note that the inner bounding arc l and the outer bounding arc n on this region both have radii equal to $r + (\delta/2)$. In calculating the holonomy the exterior angles add up to 2π, and thus, the holonomy is determined by the geodesic curvatures on the two bounding arcs. But the bottom arc l is shorter and contributes positively to the holonomy (*Why?*), and the upper arc n is longer and contributes negatively to the holonomy (*Why?*). Therefore, the holonomy (and intrinsic curvature) is negative.]

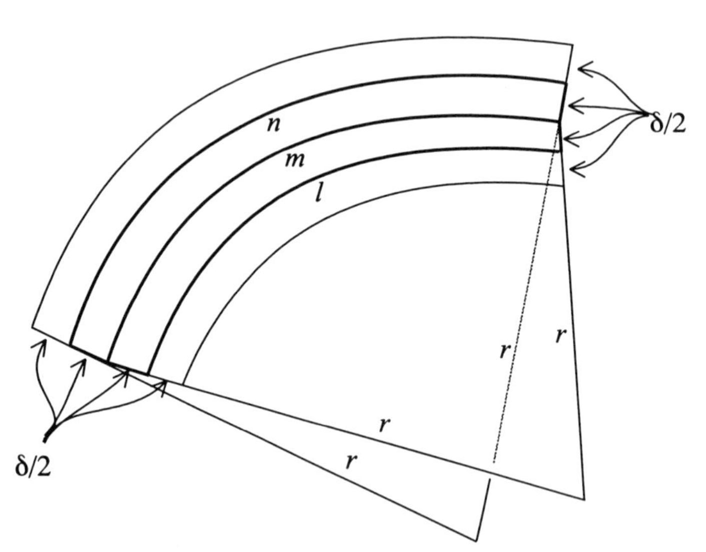

Figure 5.10. Region on the annular hyperbolic plane.

Chapter 6

Gaussian Curvature Extrinsically Defined

Pep Talk to the Reader

I think that the material in this chapter is very difficult. Don't give up and don't lose hope. What is happening here is that we are standing at the interface between things that we can see and argue about geometrically, and things that are given formally. It is difficult to hold these two aspects together—one is alive and one is dead, but both are important. It sometimes feels easier to jump headlong into the formal stuff—forgetting about what it means geometrically and just following everything through mechanically. Don't do that! Unfortunately that is often the tendency—it is also my tendency! Resist it and persevere in trying to see what the meanings of these formal things are geometrically as you go along. This is hard to do, but the effort will be well worth it. On the other hand, there exists a tendency to ignore the formal stuff and rely only on our geometric intuition. But, if we ignore the formal stuff, we would miss out on the incredibly powerful tools contained in the formalism. We need to use both the formal analytic tools and our geometric intuition; and we need to look for their interrelations. Relate everything in this chapter to the example of surfaces you already know, such as the sphere, cylinder, cone, ribbon, and strake.

In Chapter 5 we developed an intrinsic description of the intrinsic curvature of a surface. In this chapter we start with the more common extrinsic description of the Gaussian curvature of a surface, which is based on the normal curvature introduced in Problem **4.7.a**. The Gaussian and intrinsic curvatures are easily seen to be the same on a sphere. Then we use a mapping (called the ***Gauss map***) from the surface to the sphere, which then allows us to show that the Gaussian curvature and intrinsic curvature coincide on all C^2 surfaces.

In Chapter 7 we will use these results to express the Gaussian (intrinsic) curvature in local coordinates and to derive several more intrinsic descriptions of Gaussian curvature.

At the end of this chapter we will explore *mean curvature* and *minimal surfaces*.

Problem 6.1. Gaussian Curvature, Extrinsic Definition

Let p be a point on the smooth C^2 surface M in $\mathbb{R}^3$, and let $\mathbf{n}(p)$ be one of the two choices of unit normal to the surface at p, so that $\mathbf{n}$ is differentiable in a neighborhood of p. Let $\mathbf{T}_p$ be a unit tangent vector at p. If γ is a curve on M, which passes through p and has $\mathbf{T}_p$ as unit tangent vector, then, according to Problem **4.7.a**, the normal curvature of γ at p satisfies

$$\kappa_{\mathbf{n}}(p) = \langle \mathbf{T}_p, -\nabla_{\mathbf{T}_p}\mathbf{n} \rangle\, \mathbf{n}(p).$$

Since $\mathbf{n}(p)$ is a unit vector, $\langle \mathbf{T}_p, -\nabla_{\mathbf{T}_p}\mathbf{n} \rangle$ is the magnitude of the normal curvature vector, and thus we define the (***scalar***) ***normal curvature*** of M at p in the direction $\mathbf{T}_p$ as

$$\kappa_{\mathbf{n}}(\mathbf{T}_p) \equiv \langle \mathbf{T}_p, -\nabla_{\mathbf{T}_p}\mathbf{n} \rangle$$
$$\equiv \text{the length of the projection of } -\nabla_{\mathbf{T}_p}\mathbf{n} \text{ onto the direction of } \mathbf{T}_p.$$

Note that $\kappa_{\mathbf{n}}(\mathbf{T})$ is positive when $\kappa_{\mathbf{n}}$ is in the direction of $\mathbf{n}$ and negative otherwise. In Problem **4.7.a** we learned that $\kappa_{\mathbf{n}}(\mathbf{T}_p)$ is the normal curvature of *any* unit speed curve through p, which has $\mathbf{T}_p$ as velocity vector at p. Some books use the name ***Weingarten map*** to indicate the map $\mathsf{L}(\mathbf{X}_p) = -\nabla_{\mathbf{X}_p}\mathbf{n}$.

Recall from Chapter 3 that the normal curvature of a curve on the surface M is the curvature of the curve that is due to its being on the surface. So, the normal curvature $\kappa_{\mathbf{n}}(\mathbf{T}_p)$ tells us how the surface is curving in the direction of $\mathbf{T}_p$. Then $\kappa_{\mathbf{n}}$ is a real-valued function defined on the unit vectors (unit circle) in the tangent space T_pM; as such, if $\kappa_{\mathbf{n}}$ is continuous, then it has a maximum value and a minimum value, which we shall denote κ_1 and κ_2.

These are called the ***principal curvatures***, and the directions in which they occur are called the ***principal directions***. We then define the ***Gaussian curvature*** of the surface at the point p to be the product of κ_1 and κ_2. Note that this is an extrinsic definition of the Gaussian curvature. In Problem **6.4** we will show that this (extrinsically defined) Gaussian curvature coincides with the intrinsic curvature on C^2 surfaces.

a. *What are the principal directions, and principal curvatures of the cylinder, cone, and sphere? Give geometric or analytic reasons.*

If **A**, **B** are linearly independent vectors based at the point p, then the ***span of*** **A**, **B**, denoted by "span[**A**,**B**]", is the plane (through p) determined by the two vectors.

b. *Show that if* γ *is a unit speed curve in the smooth surface* M, *and*

$$\gamma^* = \{\text{span}[\gamma'(0), \mathbf{n}(\gamma(0))] \cap M\},$$

(see Figure 6.1) *then, at* $\gamma(0) = p$, *the curvature vectors on* M *satisfy*:

$$\kappa(\gamma^*) = \kappa_{\mathbf{n}}(\gamma^*) = \kappa_{\mathbf{n}}(\gamma) = \langle \gamma'(0), -\gamma'(0)\mathbf{n} \rangle\, \mathbf{n},$$

and the scalar curvatures satisfy:

$$\pm\kappa(\gamma^*) = \kappa_{\mathbf{n}}(\gamma^*) = \kappa_{\mathbf{n}}(\gamma) = \langle \gamma'(0), -\gamma'(0)\mathbf{n} \rangle,$$

where $\kappa_{\mathbf{n}}(\gamma^*) = \kappa_{\mathbf{n}}(\gamma)$ *is negative if the curves curve away from the normal* **n**, *as in Figure 6.1.*

Note also that $\kappa(\gamma^*) = \kappa_{\mathbf{n}}(\gamma^*)$ is only asserted to hold at the one point $\gamma(0)=p$. This does not imply that γ^* is a geodesic, even though $\kappa_g = 0$, because it is only zero at that one point. Look at an example of this by looking at the tangent vector to a helix along a cylinder. The corresponding γ^* on the cylinder will be an ellipse tangent to the (intrinsically straight) helix but will not coincide with it.

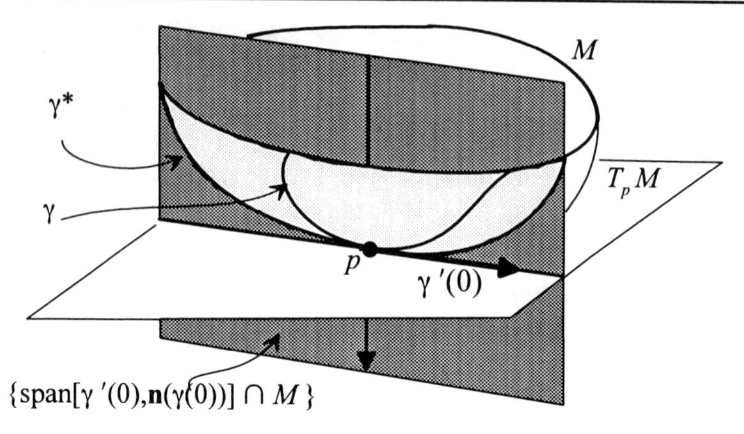

Figure 6.1. Finding the normal curvature.

***c.** *Look at the surface which is the graph of*

$$z = (1 - \cos 4\theta)\, r^2.$$

Find the principal directions and principal curvatures at (0,0,0), *and note that the principal directions are not perpendicular. Note that similar examples are possible in the general form*

$$z = f(\theta)\, r^2,$$

where f is any twice differentiable function that satisfies:

$$f(-x) = f(x).$$

Computer Exercise 6.1 will allow you to display and view these surfaces. When there is a C^2 local coordinate patch, the principal directions are orthogonal, as you shall see in the next problem. Thus it must be that *the surfaces in part* **c** *must not have any* C^2 *coordinate patch.*

***d.** *Show that every closed smooth surface* (see Problem **4.2.c**) *has at least one point at which all the normal curvatures are positive with respect to the inward pointing normal.*

[Hint: Start with a sphere that contains the surface in its interior and then gradually shrink the sphere until it first touches the surface. What can you say about the surface at this point of first touching?]

PROBLEM 6.2. Second Fundamental Form

If M is a smooth surface in $\mathbb{R}^3$, and $\mathbf{V}_p$ and $\mathbf{W}_p$ are orthogonal unit tangent vectors at a point p on M, then any other unit tangent vector $\mathbf{T}_p$ at p can be written as a linear combination: $\mathbf{T}_p = a\mathbf{V}_p + b\mathbf{W}_p$. If we are to use local coordinates then we would like to be able to calculate $\kappa_{\mathbf{n}}(\mathbf{T}_p)$, knowing only $\kappa_{\mathbf{n}}(\mathbf{V}_p)$, $\kappa_{\mathbf{n}}(\mathbf{W}_p)$ and a and b. However, $\kappa_{\mathbf{n}}$ is not a linear function. In fact:

a. *Show that:*

$$\kappa_{\mathbf{n}}(a\mathbf{V}_p + b\mathbf{W}_p) =$$

$$= a^2\,\kappa_{\mathbf{n}}(\mathbf{V}_p) + b^2\,\kappa_{\mathbf{n}}(\mathbf{W}_p) + ab\,\langle \mathbf{V}_p, -\mathbf{W}_p\mathbf{n}\rangle + ab\,\langle \mathbf{W}_p, -\mathbf{V}_p\mathbf{n}\rangle.$$

Thus, it is important to look at the quantities such as

$$\langle \mathbf{V}_p, -\mathbf{W}_p\mathbf{n}\rangle \text{ with } \mathbf{V}_p \neq \mathbf{W}_p.$$

So, if $\mathbf{X}_p, \mathbf{Y}_p \in T_pM$ for a smooth surface M in $\mathbb{R}^3$, then we define the ***second fundamental form*** to be:

$$\mathrm{II}(\mathbf{X}_p, \mathbf{Y}_p) = \langle \mathbf{X}_p, -\mathbf{Y}_p\mathbf{n}\rangle,$$

where $\mathbf{n}(q)$ is a differentiable choice of unit normal to M at all points q near p. We are interested, in the end, only in the normal curvature in any direction

$$\mathbf{T}_p\!: \kappa_{\mathbf{n}}(\mathbf{T}_p) = \mathrm{II}(\mathbf{T}_p, \mathbf{T}_p).$$

The general Second Fundamental Form and its mixed terms $\langle \mathbf{X}_p, -\mathbf{Y}_p\mathbf{n}\rangle$ will be needed only when we want to express the normal curvature in terms of local coordinates. Now,

b. *Show that the second fundamental form is bilinear (linear in each variable).*

[Hint: Use Problem **4.8**.]

Let M be a smooth surface in $\mathbb{R}^3$ with C^2 local coordinates $\mathbf{x}(u^1,u^2)$ with $\mathbf{x}(a,b) = p$.

c. *Show that*

$$\mathrm{II}(\mathbf{x}_1,\mathbf{x}_2) = \langle \mathbf{x}_{21},\mathbf{n} \rangle = \langle \mathbf{x}_{12},\mathbf{n} \rangle = \mathrm{II}(\mathbf{x}_2,\mathbf{x}_1)$$

and that

$$\mathrm{II}(\mathbf{X}_p,\mathbf{Y}_p) = \mathrm{II}(\mathbf{Y}_p,\mathbf{X}_p).$$

[Hint: Look at $\mathbf{x}_1\langle \mathbf{x}_2,\mathbf{n} \rangle = 0$, using the local coordinates $\mathbf{x}(u^1,u^2)$, and then write

$$\mathbf{X}_p = \Sigma\, X^i\mathbf{x}_i(a,b) \text{ and } \mathbf{Y}_p = \Sigma\, Y^i\mathbf{x}_i(a,b),$$

and use Problem **4.8**.]

Part **c** is the only place that the assumption C^2 is used in a crucial way in all its power (that is, C^2 requires that all first and second partial derivatives exist and are continuous *and* that the mixed partials are equal).

d. *Show that the maximum and minimum of*

$$\{\mathrm{II}(\mathbf{T},\mathbf{T}) \mid \mathbf{T} \text{ a unit tangent vector at } p\}$$

occur when $-\mathbf{Tn} = \lambda\mathbf{T}$. *Thus, in this case, the rate of change of the normal* **n** *with respect to* **T** *is in a direction parallel to* **T**.

[Hint: Note that $\mathrm{II}(-\mathbf{T},-\mathbf{T}) = \mathrm{II}(\mathbf{T},\mathbf{T})$. Part **d** may be solved in at least three ways:

1. using from **analysis** the theory of Lagrange multipliers to maximize/minimize $\mathrm{II}(\mathbf{X},\mathbf{X})$, subject to the constraint that $\langle \mathbf{X},\mathbf{X} \rangle = 1$.

2. expressing $\mathrm{II}(\mathbf{T},\mathbf{T})$ in terms of local coordinates and then using (from **linear algebra**) the theory of eigenvalues and eigenvectors of symmetric matrices or quadratic forms.

3. arguing **geometrically** that if $\mathbf{T}$ is a maximum or minimum, then $\frac{d}{dh}\text{II}(\mathbf{T}+h\mathbf{T}^{\perp},\mathbf{T}+h\mathbf{T}^{\perp})_{h=0}=0$ (where $\mathbf{T}^{\perp}$ is a unit vector perpendicular to $\mathbf{T}$) and then differentiate.]

e. *Show that, if* $\mathbf{T}_1$ *and* $\mathbf{T}_2$ *are unit tangent vectors such that*

$$-\mathbf{T}_1\mathbf{n}=\lambda_1\mathbf{T}_1 \textit{ and } -\mathbf{T}_2\mathbf{n}=\lambda_2\mathbf{T}_2,$$

then either

$$\lambda_1=\lambda_2 \textit{ or } \mathbf{T}_1 \textit{ is perpendicular to } \mathbf{T}_2.$$

[Hint: Consider $\text{II}(\mathbf{T}_1,\mathbf{T}_2)$ and remember that $\text{II}(\mathbf{T},\mathbf{T})$ is the normal curvature of the surface in the direction $\mathbf{T}$.]

It follows that if $\lambda_1\neq\lambda_2$, then $\mathbf{T}_1$ and $\mathbf{T}_2$ must be the principal directions at p. If $\lambda_1=\lambda_2$, then the normal curvature is the same in all directions and any two perpendicular unit vectors can be chosen as the principal directions. Note also that

$$\kappa_i=\text{II}(\mathbf{T}_i,\mathbf{T}_i)=\langle\mathbf{T}_i,-\mathbf{T}_i\mathbf{n}\rangle=\langle\mathbf{T}_i,\lambda_i\mathbf{T}_i\rangle=\lambda_i\langle\mathbf{T}_i,\mathbf{T}_i\rangle=\lambda_i.$$

f. *What are the principal directions on a* C^2 *surface of revolution expressed in rectangular coordinates as*

$$(r(z)\cos\theta,\ r(z)\sin\theta,\ z)?$$

Show that in these directions the principal curvatures (with respect to the inward pointing normal) are:

$$\kappa_1=\frac{-r''(z)}{\left[1+(r'(z))^2\right]^{3/2}} \textit{ and } \kappa_2=\frac{1}{r(z)\sqrt{1+(r'(z))^2}}.$$

[Hint: Argue geometrically (using **6.2.d**) which are the principal directions. Then calculate the normal curvatures of curves in those directions. Do NOT use the second fundamental form. Note that a surface of revolution is C^2 whenever $r(z)$ is positive and C^2. You may find helpful the discussion in the section *Curvature of the Graph of a Function*, preceding Problem **2.4**.]

To summarize the above discussion: The directions, $\mathbf{T}_1, \mathbf{T}_2$, in which the maximum and minimum of II(**T**,**T**) occur, are called the ***principal directions at*** p and the values of II(**T**,**T**) in these directions, κ_1, κ_2, are called ***the principal curvatures at*** p. Note that, κ_1, κ_2, are (by Problem **5.1**) the normal curvatures of unit speed curves in the principal directions. The product $\kappa_1\kappa_2$ is called the ***Gaussian curvature at*** p. The above (because it involves the unit normal to the surface) is an extrinsic description of the Gaussian curvature; but below we will show that, in fact, the Gaussian curvature is intrinsic and that it coincides with the intrinsic curvature defined in Chapter 5. In Chapter 7 we can provide an intrinsic description of Gaussian curvature in terms of local coordinates.

Thus in the principal directions, $\mathbf{T}_1, \mathbf{T}_2$, at a point p, we can write $-\mathbf{T}_i\mathbf{n} = \kappa_i\mathbf{T}_i$. In these principal directions the rate of change of the normal to the surface is equal in magnitude to normal curvature. By **6.1**, for the curve γ^*, $\kappa(\gamma^*) = \kappa_{\mathbf{n}}(\gamma^*)$ (see Figure 6.1), and thus by Problem **2.3** the normal to γ^* is parallel to the normal to the surface (at p), and their rates of change along γ^* are equal in both magnitude and direction at p. It is worthwhile spending as much time as needed to understand this situation because it will keep coming up and will be crucial later on.

PROBLEM 6.3. The Gauss Map

We now want to connect what we know from Chapter 5 about the connections between curvature and holonomy on a sphere to similar connections on an arbitrary smooth surface M in $\mathbb{R}^3$. The connection will be via the ***Gauss map*** (or ***normal spherical image***), which is a function from the surface to the unit sphere $\mathbf{S}^2$. The Gauss map is defined as follows: Start with a point p on M and choose a unit normal $\mathbf{n}(p)$ to the surface at that point. Then, if $\mathbf{n}(p)$ is considered as a vector bound at the origin, its head is on a point of the unit sphere. We define this point to be the image of p under the Gauss map. (See Figure 6.2.) We usually write the Gauss map as $\mathbf{n}: M \rightarrow \mathbf{S}^2$. There are at every point two possible unit normals, and we assume that we have to pick one of them at each point so that $\mathbf{n}$ is continuously defined.

You may find helpful Computer Exercise 6.3, which will allow the display of images of the Gauss map. But it is more important for you to argue through specific examples, such as in Part **a**, below. You probably will find it helpful to use physical models of the surface and a sphere.

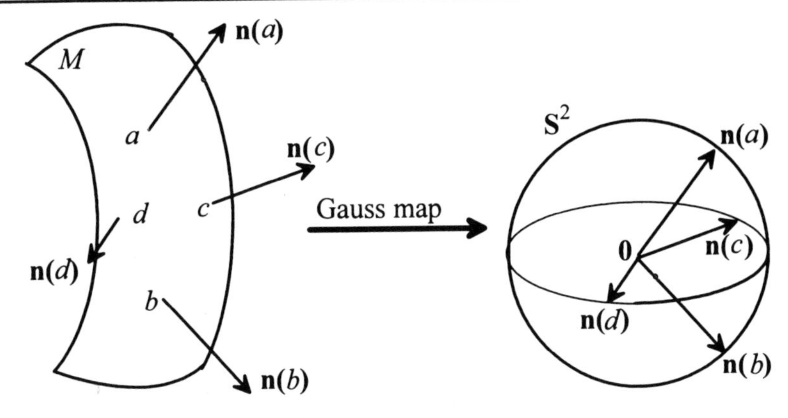

Figure 6.2. Gauss map.

a. *What is the spherical image of cylinders and spheres of different radii? What is the spherical image of a normal circular cone and how does it depend on the cone angle? Describe the spherical image of a strake. Describe the spherical image of a torus and divide the torus into eight regions each congruent to either A or B in Problem* **5.5.c** *and Figure 5.9. What do you notice as you shift from the inside saddle-shaped A-regions on the torus to the outside B-regions? As you move around the boundary curves in a counterclockwise direction how do the images on the unit sphere move?*

b. *If* $\mathbf{P}(s)$ *is a parallel vector field along the curve* $\gamma(s)$ *on the surface M, then* $\mathbf{P}(s)$ *is also a parallel vector field along the curve* $\mathbf{n}(\gamma(s))$ *on the sphere.*

[Hint: First show that if a vector $\mathbf{V}$ is tangent to the surface at $\gamma(s)$, then $\mathbf{V}$ is also tangent to the sphere at $\mathbf{n}(\gamma(s))$. Thus $\mathbf{P}(s)$ is a vector field along $\mathbf{n}(\gamma(s))$. Now look at the rate of change of $\mathbf{P}(s)$ with respect to arclength at $\gamma(s)$ and then at $\mathbf{n}(\gamma(s))$. Remember that, if s is arclength on γ then it will, in general, not be arclength on $\mathbf{n}(\gamma)$.]

c. *If* γ *is a "small" piecewise smooth closed curve on M, then* $\mathscr{H}(\gamma) = \mathscr{H}(\mathbf{n}(\gamma))$. *You must decide what "small" means.*

[Hint: Use the definition of holonomy as an angle between a vector and its parallel transport. (See Chapter 5.) You need to specify "small" in order to avoid the $n2\pi$ ambiguity in the measure of the angle.]

d. *If λ is a piecewise smooth curve on a C^2 surface M such that the velocity vector λ' is in a principal direction $\mathbf{T}_1$ with principal curvature κ_1, then the velocity vector of the curve $s \to \mathbf{n}(\lambda(s))$ is equal to $-\kappa_1 \lambda'(s)$.*

[Hint: Use Problem **6.2.d-e**.]

PROBLEM 6.4. Gauss-Bonnet and Intrinsic Curvature

a. *If R is a "small" region on a C^2 surface M with Gauss map $\mathbf{n}: M \to S^2$, then*

$$\iint_R \kappa_1\kappa_2 \, dA = \text{Area}\,(\,\mathbf{n}(R)\,).$$

[Hint: At the point p in M, choose any local C^2 coordinates $\mathbf{x}$ such that $\mathbf{x}_1$ and $\mathbf{x}_2$ are in the principal directions and use Problem **6.3.d**. You must decide what you need "small" to mean.]

b. *If R is a "small" region bounded by a piecewise smooth curve γ on a C^2 surface M, then show that*

$$\mathcal{H}(R) = 2\pi - \int_\gamma \kappa_g \, ds - \Sigma\alpha_i = \iint_R \kappa_1\kappa_2 \, dA,$$

where the double integral is the (surface) integral over R.

[Hint: Use Problem **5.4.b**.]

c. *Now, show that Gaussian curvature $\kappa_1(p)\kappa_2(p)$ at a point p on any C^2 surface is equal to the intrinsic curvature defined as:*

$$K(p) = \lim_{n\to\infty} \mathcal{H}(R_n) \,/\, A(R_n),$$

where $\{R_n\}$ is a sequence of small (geodesic) polygons that converge to p.

[Hint: Use Problem **6.2.e**.]

This leads to the famous result of Gauss, which is contained in translation in [**DG**: Gauss]. "Theorema Egregium" means "Remarkable Theorem" in Latin, the original language of Gauss' paper.

d. (Gauss' Theorema Egregium) *If*

$$f\colon M \to \mathbb{R}^3 \text{ and } g\colon M \to \mathbb{R}^3$$

are two C^2 *imbeddings that preserve the Riemannian metric, then the Gaussian curvature of* $f(M)$ *at* $f(p)$ *equals the Gaussian curvature of* $g(M)$ *at* $g(p)$.

PROBLEM 6.5. Matrix of the Second Fundamental Form

Let M be a smooth surface in $\mathbb{R}^3$ with C^2 local coordinates $\mathbf{x}(u^1,u^2)$.

a. *Show that if*

$$\mathbf{X}_p = \Sigma\, X^i \mathbf{x}_i(a,b) \text{ and } \mathbf{Y}_p = \Sigma\, Y^i \mathbf{x}_i(a,b),$$

then

$$\mathrm{II}(\mathbf{X}_p, \mathbf{Y}_p) =$$

$$= (X^1 X^2) \begin{pmatrix} \mathrm{II}(\mathbf{x}_1(a,b), \mathbf{x}_1(a,b)) & \mathrm{II}(\mathbf{x}_1(a,b), \mathbf{x}_2(a,b)) \\ \mathrm{II}(\mathbf{x}_2(a,b), \mathbf{x}_1(a,b)) & \mathrm{II}(\mathbf{x}_2(a,b), \mathbf{x}_2(a,b)) \end{pmatrix} \begin{pmatrix} Y^1 \\ Y^2 \end{pmatrix} =$$

$$= (X^1 X^2) \begin{pmatrix} \langle \mathbf{x}_{11}(a,b), \mathbf{n}(a,b) \rangle & \langle \mathbf{x}_{12}(a,b), \mathbf{n}(a,b) \rangle \\ \langle \mathbf{x}_{21}(a,b), \mathbf{n}(a,b) \rangle & \langle \mathbf{x}_{22}(a,b), \mathbf{n}(a,b) \rangle \end{pmatrix} \begin{pmatrix} Y^1 \\ Y^2 \end{pmatrix}.$$

The above matrix is called the ***matrix of the second fundamental form*** (in local coordinates $\mathbf{x}(u^1,u^2)$). Some books call the matrix simply the second fundamental form.

b. *If we choose local coordinates such that at* $p = \mathbf{x}(a,b)$ *we have*

$$\mathbf{x}_1(a,b) = \mathbf{T}_1 \text{ and } \mathbf{x}_2(a,b) = \mathbf{T}_2,$$

then show that the matrix of the second fundamental form is

$$\begin{pmatrix} \kappa_1 & 0 \\ 0 & \kappa_2 \end{pmatrix},$$

and if $\mathbf{T}(\theta)$ *denotes the unit vector that is in a direction at an angle of* θ *away from the first principal direction* $\mathbf{T}_1$, *then show that the normal curvature is given by*

$$\kappa_{\mathbf{n}}(\mathbf{T}(\theta)) = \kappa_1 \cos^2\theta + \kappa_2 \sin^2\theta.$$

Note on a sphere that the normal curvature is the same in all directions, and thus, any orthogonal local coordinates on the sphere will have their second fundamental form matrix be a diagonal matrix. This is also true for the standard local coordinates on the cylinder and cone. However, it is not true for the standard local coordinates on the strake.

***c.** *Suppose that* $\mathbf{x}$ *expresses* M *as the graph of a function*:

$$\mathbf{x}(x,y) = (x, y, f(x,y)).$$

Show that, at $p = \mathbf{x}(a,b)$,

$$\mathbf{x}_1(a,b) = (1,0,f_x(a,b)),\ \ \mathbf{x}_2(a,b) = (0,1,f_y(a,b))$$

and

$$\mathbf{n}(a,b) = \frac{\mathbf{x}_1(a,b) \times \mathbf{x}_2(a,b)}{|\mathbf{x}_1(a,b) \times \mathbf{x}_2(a,b)|} = \frac{(-f_x, -f_y, 1)}{\sqrt{1 + (f_x)^2 + (f_y)^2}},$$

where

$$f_x = \frac{\partial}{\partial x} f(x,b)_{x=a} \text{ and } f_y = \frac{\partial}{\partial y} f(a,y)_{y=b}.$$

Find the matrix of the second fundamental form in these local coordinates at $\mathbf{x}(0,0)$.

[Note that the tangent vectors $\mathbf{x}_1$ and $\mathbf{x}_2$ are *not* partial derivatives of f.]

*PROBLEM 6.6. Mean Curvature and Minimal Surfaces

Using Problem **6.5.b**, we can calculate (using first year calculus) the ***mean curvature of** M **at** p*:

$$H = \frac{1}{2\pi}\int_0^{2\pi}(\mathbf{T}(\theta))d\theta = \frac{1}{2\pi}\int_0^{2\pi}[\kappa_1\cos^2\theta + \kappa_2\sin^2\theta]\,d\theta = \frac{1}{2}(\kappa_1 + \kappa_2).$$

[Note: Some texts *define* the "mean" curvature as $\kappa_1 + \kappa_2$, but this goes against the meaning of "mean" as "average."]

If the mean curvature is zero, then, either

$$\kappa_1 = 0 = \kappa_2 \text{ and } K = 0,$$

as in the case of the plane, or

$$\kappa_1 = -\kappa_2 \text{ and } K = -(\kappa_1)^2.$$

a. *Show geometrically (or by directly calculating) that the helixes that spiral up the strake and the horizontal lines on the strake all have zero normal curvature in the strake. Use this to show that the strake has zero mean curvature H.*

[Hint: The curves with zero normal curvature on the strake are not in the principal directions. Use Problem **6.5.b**.]

b. *Show that an element of area dA on the surface that is pushed in the direction of the normal will have its area change at the rate of* $-2H\,dA$.

[Hint: To get a feel for this, first show it directly for the sphere and cylinder by expressing their areas in terms of the radius r, and then (since the normal is in the direction of r) finding the rate of change of the areas by differentiating with respect to r. Use local orthonormal coordinates (x, y) in the principal directions. In each of the principal directions, draw a picture of the osculating circle with radius of 1/(normal curvature). Then we have the picture in Figure 6.3. In this picture you can see that the derivative

$$\frac{d}{dh}l_h = -\kappa\,dx.$$

Then set $dA = dx\,dy$ and let $A(h)$ be the area after dA is pushed a distance h in the direction of $\mathbf{n}$. Find $\frac{d}{dh}A$.]

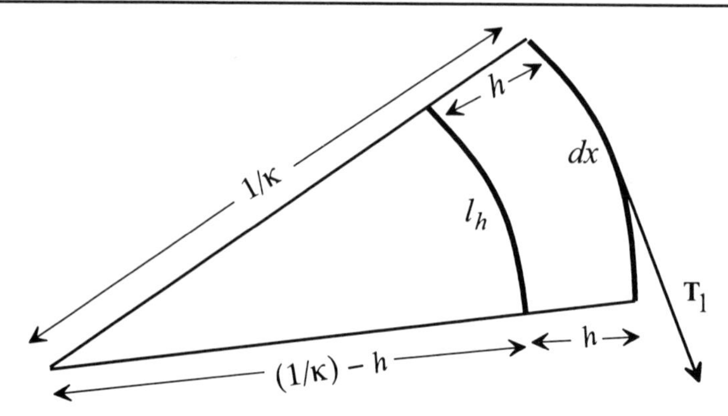

Figure 6.3. Change of arclength in the direction of the curvature.

Thus, a perturbation of a surface with zero mean curvature does not change its area. Traditionally, a surface M with zero mean curvature is called a ***minimal surface***. A soap film with equal pressures on both sides is an example of a minimal surface.

c. *Show that the surface of revolution*

$$\mathbf{x}(\theta,z) = ((1/a)\cosh(az+b)\cos\theta, (1/a)\cosh(az+b)\sin\theta, z)$$

is a minimal surface. This surface is called a ***catenoid***.

[Hint: Use Problem **6.2.f**. As a reminder: $\cosh(x) = ½ (e^x + e^{-x})$.]

d. *Show that the catenoids are the only surfaces of revolution,*

$$(r(z)\cos\theta, r(z)\sin\theta, z),$$

which are minimal surfaces.

[Hint: Find a (second order, nonlinear) differential equation that $r(z)$ must satisfy in order for a surface of revolution

$$(r(z)\cos\theta, r(z)\sin\theta, z)$$

to be a minimal surface. Then use the fact that this differential equation has a unique solution for given initial conditions.]

Note that the plane is a minimal surface and can also be considered as a surface of revolution, but it is not of the form

$$(r(z)\cos\theta, r(z)\sin\theta, z).$$

e. *Show that the catenoid and the helicoid are locally isometric.*

[Hint: Express both the catenoid and the helicoid in geodesic rectangular coordinates. For the catenoid, set $b = 0$ and use the circle $z = 0$ (minus a point) as the base curve. For the helicoid, use the center line (the z-axis) as the base curve. Then use the result of Problem **4.9** to express the respective Riemannian metrics.]

Computer Exercise 6.6 will allow you to display and observe a transition from the helicoid to the catenoid.

For more discussion and further bibliography about minimal surfaces see [**Mi**: Osserman(1986)], [**Mi**: Osserman(1989)], and [**Mi**: Morgan].

Celebration of Our Hard Work

You have just traversed some difficult territory that mathematical pioneers struggled with for about 80 years. The results in Problems **6.1**, **6.2** and **6.5** were mostly proved by Euler in 1760. Note that these define curvature (which of course was not yet called *Gaussian* curvature) extrinsically. There seemed to be no suspicion that the curvature could be intrinsic. Thus, when Gauss first discovered this fact, he called the result "egregium Theorema," "remarkable Theorem" in Latin, the original language of Gauss' paper.

The results in Problems **6.3** and **6.4** were developed by Gauss before 1827. In [**DG**: Gauss] Gauss derived his Theorema Egregium as a corollary of results similar to those in Problem **7.1**, using local coordinates. However, there is evidence in unpublished papers (which are also included in translation in [**DG**: Gauss]) that he originally arrived at this result in much the same way we do in Problem **6.4**.

The theory of minimal surfaces dates back at least to Euler in 1744, and the results in Problem **6.6** were mostly known through the work of Lagrange and Meusnier before 1785. However, research on minimal surfaces in $\mathbb{R}^3$ is still active, (see [**Mi**: Hoffman] for a description of the discovery of a new minimal surface in 1987).

We now have under our control powerful ideas that combine the intuitive geometric ideas with formal analytic ideas. In Chapter 7 we will use our new knowledge and power to derive a number of applications of Gaussian curvature and in the process find other intrinsic descriptions.

Chapter 7

Applications of Gaussian Curvature

In this chapter we will use the hard won result from Chapter 6 to express Gaussian (intrinsic) curvature in local coordinates and to find several intrinsic descriptions of Gaussian curvature. Along the way, we will investigate the exponential map and finally come to some resolution concerning the tension between *shortest* and *straight*.

PROBLEM 7.1. Gaussian Curvature in Local Coordinates

In local coordinates $\mathbf{x}$, the second fundamental form

$$\mathrm{II}(\mathbf{X},\mathbf{Y}) = \mathrm{II}(X^1\mathbf{x}_1+X^2\mathbf{x}_2,\ Y^1\mathbf{x}_1+Y^2\mathbf{x}_2)$$

can be written as:

$$\mathrm{II}\left(\begin{pmatrix} X^1 \\ X^2 \end{pmatrix}, \begin{pmatrix} Y^1 \\ Y^2 \end{pmatrix}\right) = (X^1\ X^2)\begin{pmatrix} \langle \mathbf{x}_{11},\mathbf{n}\rangle & \langle \mathbf{x}_{12},\mathbf{n}\rangle \\ \langle \mathbf{x}_{21},\mathbf{n}\rangle & \langle \mathbf{x}_{22},\mathbf{n}\rangle \end{pmatrix}\begin{pmatrix} Y^1 \\ Y^2 \end{pmatrix}.$$

Now express the **unit** principal directions in these coordinates:

$$\mathbf{T}_1 = \begin{pmatrix} T_1^1 \\ T_1^2 \end{pmatrix} \text{ and } \mathbf{T}_2 = \begin{pmatrix} T_2^1 \\ T_2^2 \end{pmatrix}.$$

Since (from Problem **6.2**),

$$\nabla_{\mathbf{T}_1}\mathbf{n} = -\kappa_1\mathbf{T}_1 \text{ and } \nabla_{\mathbf{T}_2}\mathbf{n} = -\kappa_2\mathbf{T}_2,$$

we have

$$\mathrm{II}(\mathbf{T}_1,\mathbf{T}_1) = \kappa_1,\ \mathrm{II}(\mathbf{T}_2,\mathbf{T}_2) = \kappa_2, \text{ and } \mathrm{II}(\mathbf{T}_1,\mathbf{T}_2) = \mathrm{II}(\mathbf{T}_2,\mathbf{T}_1) = 0.$$

Thus, we can see that:

$$\begin{pmatrix} T_1^1 & T_1^2 \\ T_2^1 & T_2^2 \end{pmatrix}\begin{pmatrix} \langle \mathbf{x}_{11},\mathbf{n}\rangle & \langle \mathbf{x}_{12},\mathbf{n}\rangle \\ \langle \mathbf{x}_{21},\mathbf{n}\rangle & \langle \mathbf{x}_{22},\mathbf{n}\rangle \end{pmatrix}\begin{pmatrix} T_1^1 & T_2^1 \\ T_1^2 & T_2^2 \end{pmatrix} = \begin{pmatrix} \kappa_1 & 0 \\ 0 & \kappa_2 \end{pmatrix}.$$

Thus (using the result from matrix algebra that the determinant of a product is the product of the determinants):

$$K = \kappa_1\kappa_2 = \det\begin{pmatrix} \kappa_1 & 0 \\ 0 & \kappa_2 \end{pmatrix} =$$

$$= \det\begin{pmatrix} T_1^1 & T_1^2 \\ T_2^1 & T_2^2 \end{pmatrix}\det\begin{pmatrix} \langle \mathbf{x}_{11},\mathbf{n}\rangle & \langle \mathbf{x}_{12},\mathbf{n}\rangle \\ \langle \mathbf{x}_{21},\mathbf{n}\rangle & \langle \mathbf{x}_{22},\mathbf{n}\rangle \end{pmatrix}\det\begin{pmatrix} T_1^1 & T_2^1 \\ T_1^2 & T_2^2 \end{pmatrix}.$$

But also,

$$\langle \mathbf{T}_1,\mathbf{T}_1\rangle = 1 = \langle \mathbf{T}_2,\mathbf{T}_2\rangle \text{ and } \langle \mathbf{T}_1,\mathbf{T}_2\rangle = 0 = \langle \mathbf{T}_2,\mathbf{T}_1\rangle$$

and thus:

$$\begin{pmatrix} T_1^1 & T_1^2 \\ T_2^1 & T_2^2 \end{pmatrix}\begin{pmatrix} \langle \mathbf{x}_1,\mathbf{x}_1\rangle & \langle \mathbf{x}_1,\mathbf{x}_2\rangle \\ \langle \mathbf{x}_2,\mathbf{x}_1\rangle & \langle \mathbf{x}_2,\mathbf{x}_2\rangle \end{pmatrix}\begin{pmatrix} T_1^1 & T_2^1 \\ T_1^2 & T_2^2 \end{pmatrix} =$$

$$= \begin{pmatrix} T_1^1 & T_1^2 \\ T_2^1 & T_2^2 \end{pmatrix}(g_{ij})\begin{pmatrix} T_1^1 & T_2^1 \\ T_1^2 & T_2^2 \end{pmatrix} = \begin{pmatrix} 1 & 0 \\ 0 & 1 \end{pmatrix};$$

and

$$\det\begin{pmatrix} T_1^1 & T_1^2 \\ T_2^1 & T_2^2 \end{pmatrix}\det\begin{pmatrix} T_1^1 & T_2^1 \\ T_1^2 & T_2^2 \end{pmatrix} = (\det(g_{ij}))^{-1}.$$

Therefore, we conclude:

Theorem. *For any local coordinates* $\mathbf{x}$ *the Gaussian curvature is given by*

$$K = \kappa_1\kappa_2 = (\det(g_{ij}))^{-1}\det\begin{pmatrix} \langle \mathbf{x}_{11},\mathbf{n}\rangle & \langle \mathbf{x}_{12},\mathbf{n}\rangle \\ \langle \mathbf{x}_{21},\mathbf{n}\rangle & \langle \mathbf{x}_{22},\mathbf{n}\rangle \end{pmatrix}.$$

***a.** *If the surface has a Monge patch, show that it is the graph of a function* $(x,y,f(x,y))$ *such that the surface is tangent to the* (x,y)*-plane at* $(0,0,f(0,0))$

$$[\text{thus}, f(0,0) = 0 = f_x(0,0) = f_y(0,0)],$$

then show that

$$K = f_{xx}f_{yy} - (f_{xy})^2, \text{ at } p = (0,0,f(0,0)).$$

Warning: This formula does not hold, away from the point p, and this formula is also still extrinsic.

b. *If* $\mathbf{x}(u^1,u^2)$ *is any local coordinates with Riemannian metric matrix*

$$(g_{ij}) = \begin{pmatrix} h^2 & 0 \\ 0 & 1 \end{pmatrix},$$

where

$$h(u^1,u^2) = |\mathbf{x}_1(u^1,u^2)| \text{ and } h^2 = \langle \mathbf{x}_1, \mathbf{x}_1 \rangle,$$

and such that the second coordinate curves $g(s) \equiv \mathbf{x}(u^1,s)$ *are geodesics (parametrized by arclength, then show that the Gaussian curvature at the point* $\mathbf{x}(a,b)$ *is given by*

$$K = -\frac{h_{22}(a,b)}{h(a,b)} = -\frac{1}{h(a,b)}\left[\frac{\partial}{\partial u^2}\left(\frac{\partial}{\partial u^2}h(a,u^2)\right)\right]_{u^2=b}.$$

Note that h and its derivatives are intrinsic and thus K is intrinsic. (Note that geodesic rectangular coordinates and geodesic polar coordinates both satisfy the hypotheses of **7.1.b**, and so do the standard local coordinates on the strake.)

Outline of a proof of **7.1.b**:

1. Since

$$h = |\mathbf{x}_1| = \sqrt{\langle \mathbf{x}_1, \mathbf{x}_1 \rangle},$$

we can calculate that

$$h_2 = \frac{\langle \mathbf{x}_{21}, \mathbf{x}_1 \rangle}{\sqrt{\langle \mathbf{x}_1, \mathbf{x}_1 \rangle}}$$

and that

$$-\frac{h_{22}}{h} = -\frac{(\langle \mathbf{x}_{221}, \mathbf{x}_1 \rangle + \langle \mathbf{x}_{21}, \mathbf{x}_{21} \rangle)\langle \mathbf{x}_1, \mathbf{x}_1 \rangle - \langle \mathbf{x}_{21}, \mathbf{x}_1 \rangle^2}{\langle \mathbf{x}_1, \mathbf{x}_1 \rangle^2}.$$

2. Show that

$$\langle \mathbf{x}_{221}, \mathbf{x}_1 \rangle = \langle \mathbf{x}_{122}, \mathbf{x}_1 \rangle =$$
$$= \mathbf{x}_1 \langle \mathbf{x}_{22}, \mathbf{x}_1 \rangle - \langle \mathbf{x}_{22}, \mathbf{x}_{11} \rangle = 0 - \langle \mathbf{x}_{22}, \mathbf{x}_{11} \rangle,$$

   where the "0" results because $\mathbf{x}_{22}(a,b)$ is the curvature of the curve $\gamma(s) = \mathbf{x}(a,s)$ at $s=b$ (*Why?*) and thus is perpendicular to $\mathbf{x}_1$.

3. We can then calculate that:

$$-\frac{h_{22}}{h} = \frac{\langle \mathbf{x}_{22}, \mathbf{x}_{11} \rangle}{\langle \mathbf{x}_1, \mathbf{x}_1 \rangle} - \frac{\langle \mathbf{x}_{21}, \mathbf{x}_{12} \rangle \langle \mathbf{x}_1, \mathbf{x}_1 \rangle - \langle \mathbf{x}_{21}, \mathbf{x}_1 \rangle^2}{\langle \mathbf{x}_1, \mathbf{x}_1 \rangle^2} =$$
$$= \frac{\langle \mathbf{x}_{22}, \mathbf{x}_{11} \rangle}{\langle \mathbf{x}_1, \mathbf{x}_1 \rangle} - \frac{|\mathbf{x}_{21}|^2 (1 - \cos^2\theta)}{\langle \mathbf{x}_1, \mathbf{x}_1 \rangle},$$

   where θ is the angle from $\mathbf{x}_1$ to $\mathbf{x}_{21}$.

4. At the same time

$$K = (\det(g_{ij}))^{-1} \times \det \begin{pmatrix} \langle \mathbf{x}_{11}, \mathbf{n} \rangle & \langle \mathbf{x}_{12}, \mathbf{n} \rangle \\ \langle \mathbf{x}_{21}, \mathbf{n} \rangle & \langle \mathbf{x}_{22}, \mathbf{n} \rangle \end{pmatrix} =$$
$$= \frac{\langle \mathbf{x}_{11}, \mathbf{n} \rangle \langle \mathbf{x}_{22}, \mathbf{n} \rangle - \langle \mathbf{x}_{12}, \mathbf{n} \rangle^2}{\langle \mathbf{x}_1, \mathbf{x}_1 \rangle} =$$
$$= \frac{\langle \mathbf{x}_{22}, \mathbf{x}_{11} \rangle}{\langle \mathbf{x}_1, \mathbf{x}_1 \rangle} - \frac{|\mathbf{x}_{21}|^2 \cos^2\phi}{\langle \mathbf{x}_1, \mathbf{x}_1 \rangle},$$

   where, since $\mathbf{x}_{22}$ is in the same direction as the normal $\mathbf{n}$,

$$\langle \mathbf{x}_{11}, \mathbf{n} \rangle \langle \mathbf{x}_{22}, \mathbf{n} \rangle = \langle \mathbf{x}_{11}, \mathbf{x}_{22} \rangle,$$

   and where ϕ is the angle from $\mathbf{x}_{21}$ to $\mathbf{n}$.

5. But $\mathbf{x}_{21}$ lies in the plane of $\mathbf{x}_1$ and $\mathbf{n}$. Therefore,

$$\theta + \phi = (\text{the angle from } \mathbf{x}_1 \text{ to } \mathbf{n}) = \pi/2 \text{ and } \cos\phi = \sin\theta,$$

   and the above expressions imply that the Gaussian curvature is given by

$$K = -\frac{h_{22}}{h} \quad .$$

***c.** *For geodesic rectangular coordinates* $\mathbf{x}(u^1,u^2)$ *with base a geodesic (parametrized by arclength), show that the function*

$$f(t) = h(u^1,t) = |\mathbf{x}_1(u^1,t)|$$

satisfies for each u^1*:*

$$f(0) = 1 \ \ and \ f'(0) = \langle \mathbf{x}_{12}, \mathbf{x}_1 \rangle = -\langle \mathbf{x}_2, \mathbf{x}_{11} \rangle = 0.$$

Thus, show that, for each $u^1, f(t)$ *has a local maximum at* $t = 0$ *when* $K > 0$ *and a local minimum at* $t = 0$ *when* $K < 0$.

[Hint: Use first semester calculus.]

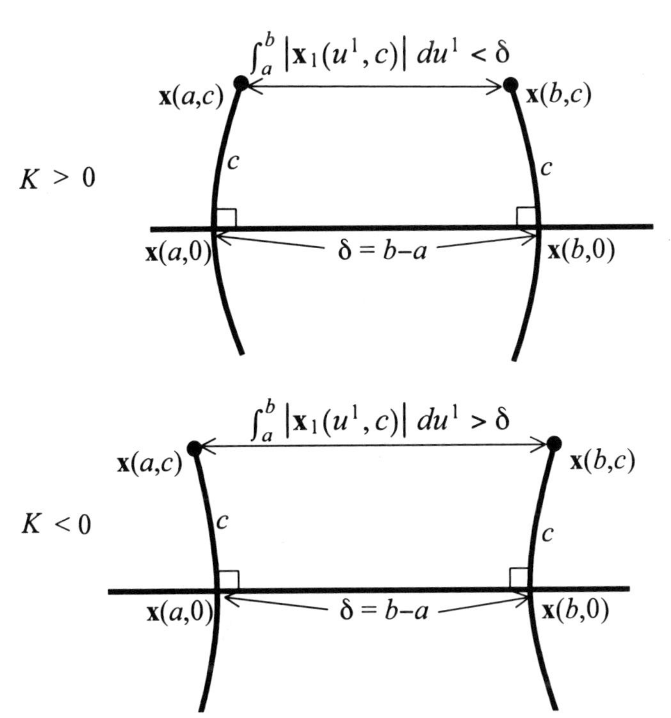

Figure 7.1. Rectangular geodesic coordinates with nonzero Gaussian curvature.

Thus, in a region in which the tangent plane is indistinguishable from the surface, h will appear to be constantly 1, and we will not be able to intrinsically determine K within such a region. However, at a distance c from the base geodesic, we have the pictures in Figure 7.1.

The results in Problem **7.1** were basically known to Gauss in 1827 and were used as the basis for his published proof [**DG**: Gauss] of the Theorema Egregium (Problem **6.4.d**).

*PROBLEM 7.2. Curvature on Sphere, Strake, Catenoid

a. *Check that the formula*

$$K = -\frac{h_{22}}{h}$$

from Problem **7.1.b** *on a sphere of radius* R *gives* $K = 1/R^2$. *Does the formula give zero curvature on cylinders and cones?*

[Hint: Use geodesic rectangular coordinates on the sphere, that is, the equator and the longitude must be parametrized by arclength, $u^1 = R\theta$ and $u^2 = R\phi$.]

b. *Calculate the Gaussian curvature for points on a strake.*

c. *Calculate the Gaussian curvature for points on a catenoid and helicoid.* (See Problem **6.6.e**.)

d. *On a sphere of radius R, show that the circumference of a circle that has intrinsic radius r can be expressed by*

$$\textit{Circumference} = 2\pi\, R \sin(\, r/R\,)$$

or, as a power series in $r^2/R^2 = r^2K$,

$$\textit{Circumference} = 2\pi r\,(1 - r^2K(1/6) + r^4K^2(\ ...\ )).$$

e. *On a sphere of radius R, show that the area of a circular disk that has intrinsic radius r is given by*

$$\textit{Area} = 2\pi\, R^2\,(1 - \cos(\, r/R\,)) =$$
$$= \pi\, r^2\,(1 - r^2K(1/12) + r^4K^2(\ ...\ )).$$

In the next problem we will use geodesic polar coordinates to show that the power series in **7.2.d-e** hold also on any C^2 surface and thus can be used to find more intrinsic descriptions of Gaussian curvature.

PROBLEM 7.3. Circles, Polar Coordinates, and Curvature

Let M be a C^2 surface. An ***intrinsic circle*** (or ***geodesic circle***) in M with radius a and center at p is the collection of all points in M that lie at a distance a along a geodesic from p. If $\mathbf{y}(\theta,r)$ ($\mathbf{y}(0,0)=p$) is geodesic polar coordinates around p, then the intrinsic circle with radius a is just the points of the form $\mathbf{y}(\theta,a)$.

If a is too large, then the circle may be distorted in various ways. For example, the intrinsic circles of radius πR on a sphere of radius R are just a point. However, for a small enough, the intrinsic circle will have a well defined area and circumference. Our goal is to find expressions for the area and circumference that are analogous to those in **7.2.d-e**.

To calculate arclength and area we first need an expression for the Riemannian metric in geodesic polar coordinates. In Problem **4.9** we showed that

$$g_{ij} = \begin{pmatrix} h^2 & 0 \\ 0 & 1 \end{pmatrix}, \text{ where } h(\theta,r) = |\mathbf{y}_1(\theta,r)|.$$

Now you can

a. *Show that, for fixed* θ, *the third Taylor approximation for h is*

$$h(\theta,r) = r - \frac{K(p)r^3}{6} + R(\theta,r), \text{ where } \lim_{r\to 0}\frac{R(\theta,r)}{r^3} = 0 .$$

[Hint: If you have forgotten about Taylor polynomials (remember, "polynomials," not "series"), then read about it in your favorite calculus text. If $f(r) \equiv h(\theta,r)$, then you can find $f(0)$ by looking directly at its definition, and $f'(0)$ you can calculate by zooming in on p sufficiently and expanding the two derivative in $f'(0)$ to their definitions. Problem **7.1.b** gives information about $f''(r)$, then take the limit. Find $f'''(0)$ by differentiating the result from **7.1.b** and then (carefully) taking the limit. That the remainder term over r^3 goes to zero, follows from the theory of Taylor polynomials, or can be checked directly and will submit to several applications of L'Hôpital's Rule.]

b. *If $C(r)$ is the circumference of a geodesic circle of (intrinsic) radius r with center at the point p, then show that*

$$C(r) = 2\pi r\,(1 - r^2K_p(1/6)) + R_C(r),$$

where $\lim_{r\to 0}(1/r^3)R_C(r) = 0$ *and* K_p *is the Gaussian curvature of the surface at* p*. Thus, we conclude*

$$K_p = \lim_{r\to 0} 3\frac{2\pi r - C(r)}{\pi r^3}.$$

[Hint: Integrate (see Problem **4.5**) and use **7.1.a**.]

c. *If $A(r)$ is the area of a geodesic circle of (intrinsic) radius r with center at the point p, then show that*

$$A(r) = \pi\, r^2\,(1 - r^2K_p(1/12)) + R_A(r),$$

where $\lim_{r\to 0}(1/r^4)R_A(r) = 0$. *Thus, we conclude*

$$K_p = \lim_{r\to 0} 12\frac{\pi r^2 - A(r)}{\pi r^4}.$$

[Hint: Integrate (see Problem **4.5**) and use **7.1.a**.]

Notice that the expressions in **7.3.b** and **7.3.c** are additional intrinsic descriptions of Gaussian curvature. In [**DG**: Spivak] these results are attributed to Diquet, Bertrand, and Puiseux in 1848.

Problem 7.4. Exponential Map and Shortest Is Straight

We now return to the issue of the connections between shortest and straight that we first encountered in Problem **1.3**. Recall that we saw then that straight paths were not always the shortest distance between their endpoints and that on the cone with cone angle 450°, even locally (near the cone point) the shortest paths were not straight. First, we investigate all the geodesics which emanate from a single point p.

Let M be a C^2 surface and T_pM be the tangent space (plane) at the point p in M. If $\mathbf{V} \in T_pM$ is a tangent vector, then there is a geodesic

$$\gamma\colon [0,1] \longrightarrow M, \text{ with } \gamma(0) = p \text{ and } \gamma'(0) = \mathbf{V},$$

and we define the ***exponential*** of **V** to be

$$\exp(\mathbf{V}) = \exp_p(\mathbf{V}) = \gamma(1).$$

The name "exponential" comes from the form it takes on Lie groups or spaces of matrices, see [**DG**: Spivak], Volume 1, Chapter 10. That the exponential is a C^2 map in some neighborhood of p follows from standard theorems about the solutions of differential equations varying smoothly with respect to their initial conditions (see [**DG**: Spivak], Volume 1, Chapter 5, for a detailed discussion).

a. *Show that, if* $\mathbf{U}(\theta)$ *is the unit vector in the direction* θ, *then the function defined by*

$$\mathbf{y}(\theta,r) = \exp(r\mathbf{U}(\theta))$$

is geodesic polar coordinates in some open neighborhood U_p *of* p. *Thus, conclude that all the geodesics in* U_p *that pass through* p *are perpendicular to the level curves*

$$\{\ \exp_p(\mathbf{V})\ |\ |\mathbf{V}| = \text{constant}\ \}.$$

[Hint: Use Problem **4.9**.]

b. *Show that any geodesic* γ *in* U_p *that joins* p *to* p^* *is the shortest path joining* p *to* p^*. (This was apparently first proved by J.H.C. Whitehead in 1932.)

Outline of a proof of **7.4.b**:

1. Assume that there is a piecewise smooth path α: $[0,b] \to U_p$ from p to p^* that is shorter than γ. Then, using geodesic polar coordinates $\mathbf{y}(\theta,r)$ we can write $\alpha(t) = \mathbf{y}(\theta(t),r(t))$. Differentiate and show that, for $0 < a \le t \le b$,

$$|\alpha'(t)| \ge |r'(t)|,$$

with equality if and only if $\theta'(t) = 0$.

2. Then integrate and show that

$$\int_a^b |\alpha'(t)|dt \ge |\ r(a) - r(b)\ |,$$

with equality if and only if $r(t)$ is monotone and $\theta(t)$ is constant.

3. Take the limit as $a \to 0$ and conclude the desired result.

c. *Let $C(a)$ be a circle of radius a and center p. Let λ be a path that joins two points, p^*, p^{**}, on $C(a)$ and which is the union of two geodesic pieces, which form an angle at p of angle $\phi<\pi$ as in Figure 7.2. Show that, if a is sufficiently small, there is a path α in the interior of $C(a)$ that has length less than $2a$ (= the length of λ).*

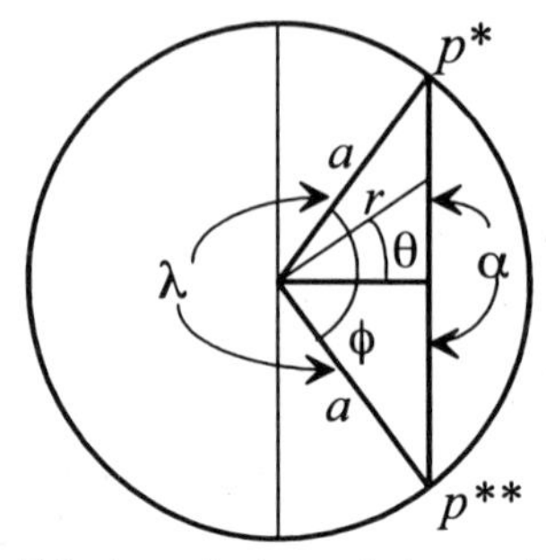

Figure 7.2. A geodesic angle is not shortest.

Outline of a proof of **7.4.c**:

1. Look at the path α marked in the figure with parametrization

$$\alpha(\theta) = \mathbf{y}(\theta, r(\theta)),\ -\phi/2 \le \theta \le \phi/2$$

where

$$r(\theta) = a\frac{\cos\phi/2}{\cos\theta}.$$

2. Look at the integral that expresses the length of α:

$$\int_{-\phi/2}^{\phi/2} |\alpha'(\theta)| d\theta,$$

and use the estimate in **7.3.a** when expanding the integrand.

3. Show that

$$\int_{-\phi/2}^{\phi/2} |\alpha'(\theta)|\, d\theta = \int_{-\phi/2}^{\phi/2} \frac{a\cos\phi/2}{\cos^2\theta}\sqrt{1 + a(A(a,\theta)) + \frac{R}{a^2}(B(a,\theta))}\, d\theta$$

where $A(a,\theta)$ and $B(a,\theta)$ are bounded for $-\phi/2 \le \theta \le \phi/2$ and $0 < a \le 1$. Thus, for sufficiently small a,

$$\sqrt{1 + a(A(a,\theta)) + \frac{R}{a^2}(B(a,\theta))} \le C < \frac{1}{\sin\phi/2},$$

and

$$\int_{-\phi/2}^{\phi/2} |\alpha'(\theta)|\, d\theta \le \int_{-\phi/2}^{\phi/2} \frac{a\cos\phi/2}{\cos^2\theta} C\, d\theta = 2aC\sin\phi/2 < 2a.$$

For the main result of this problem we need a notion of *completeness*:

> M is ***geodesically complete*** if every geodesic in M can be extended indefinitely. This is a direct interpretation of Euclid's first postulate which says: "Every straight line can be extended indefinitely."

Now we can prove the main result of this problem, which (together with **7.4.e**) is usually called the Hopf-Rinow-de Rham Theorem (proved in 1931):

d. *If M is **geodesically complete**, then any two points can be joined by a geodesic that is the shortest path between them.*

Outline of a proof of **7.4.d**:

1. Let p, q be any two points in M with their distance $d(p,q) = b$. Let C be a circle of radius δ and center p so that $C \subset U_p$. There is a point p^* on C such that

$$d(p^*,q) \le d(x,q), \text{ for all } x \in C.$$

Now $p^* = \exp_p(\delta\mathbf{V})$, for some unit tangent vector $\mathbf{V} \in T_pM$.

$$\text{CLAIM: } \exp_p(b\mathbf{V}) = q;$$

this will show that the geodesic $\gamma(t) = \exp_p(t\mathbf{V})$ is a geodesic of length b joining p to q.

2. The claim will be true (*Why?*) if

$$b \in A \equiv \{\, t \mid d(\gamma(t),q) = b - t \,\}.$$

3. Since every curve form p to q must cross C, we have

$$d(p,q) = \min_{x \in C}[d(p,x) + d(x,q)] = \delta + d(p*,q).$$

So $d(p^*,q) = b - \delta$ and $\delta \in A$.

4. Let t^* be the least upper bound of all t in A. Then $t^* \in A$. Suppose that $t^* < b$. Let C^* be the circle of radius δ^* around $\gamma(t^*)$, and let q^* be the point on C^* that is closest to q. (See Figure 7.3.)

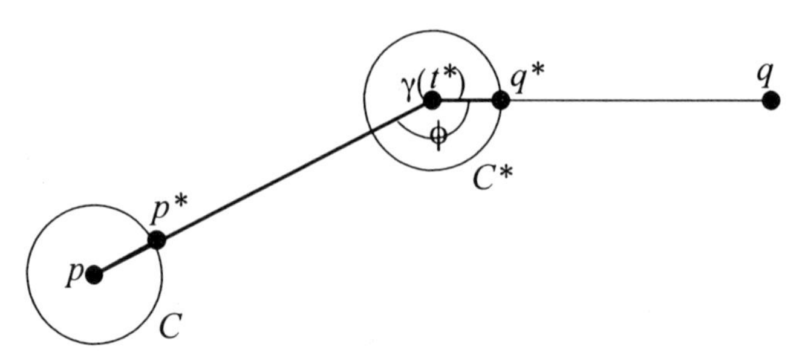

Figure 7.3. Finding shortest path from *p* to *q*.

5. Use **7.4.c** to argue that the angle ϕ in Figure 7.3 must be equal to π and thus t^* is not the least upper bound of elements in A.

There is another notion of completeness that is familiar from analysis:

> M is ***Cauchy complete*** (or, simply, ***complete***) if every Cauchy sequence in M converges. A sequence $\{x_i\}$ is a Cauchy sequence if, for every integer m, there is an integer n such that
>
> $$|x_i - x_j| < 1/m, \text{ whenever } i > n \text{ and } j > n.$$

Notice that Cauchy completeness is a local concept while geodesic completeness is a more global notion. Nevertheless, you can prove that:

***e.** *A surface M is Cauchy complete if and only if it is geodesically complete.*

[Hint: Work locally and use the results above, including that the exponential map is continuous.]

PROBLEM 7.5. Surfaces with Constant Curvature

a. *Let* $\mathbf{x}(u^1,u^2)$ *be a geodesic rectangular coordinate chart (with base curve a geodesic) on a surface M with constant Gaussian curvature K. Show that the Riemannian metric matrix is*

$$(g_{ij}) = \begin{pmatrix} h^2 & 0 \\ 0 & 1 \end{pmatrix},$$

where

$$h = h(u^1,u^2) = \begin{cases} \cos\sqrt{K}\,u^2, & \textit{if } K \geq 0 \\ \cosh\sqrt{|K|}\,u^2, & \textit{if } K \leq 0 \end{cases}.$$

[Hint: Use Problems **4.9**, **7.1.b,** and **7.1.c** and ordinary calculus.]

b. *Prove that any two surfaces with the same constant Gaussian curvature are locally isometric.*

[Hint: Use geodesic rectangular coordinates (with the base curve a geodesic) on both surfaces and define a map that takes a point on the first surface to the point on the other surface with the same coordinates. With respect to these coordinates, show that the Riemannian metrics of the two surfaces are equal. The arclength of any curve γ on a surface is given by the integral

$$\int |\gamma'(t)|\, dt = \int \sqrt{\langle \gamma'(t), \gamma'(t) \rangle}\, dt.$$

Thus, if the Riemannian metrics are the same, then all lengths are the same on the two surfaces.]

c. *Show that on a surface of constant curvature there exist locally: rotations about any point through any angle, translations along any geodesic, and reflections across any geodesic.*

[Hint: Apply your argument in Part **b** to different local geodesic rectangular coordinates on the same surface.]

Since the annular hyperbolic plane is constructed the same everywhere (as $\delta \to 0$), it is homogeneous (that is, intrinsically and geometrically every point has a neighborhood that is isometric to a neighborhood of any other point). Thus the Gaussian curvature is constant. In addition (if the construction is continued indefinitely), every geodesic can be continued indefinitely in both directions. Thus:

> *The annular hyperbolic plane has global translations, rotations, and reflections, that are isometries of the whole surface onto itself.*

Problem 7.6. Ruled Surfaces and Ribbons

Now we are in a position to finish some details about ruled surfaces and the converse of the Ribbon Test, as was promised at the end of Chapter 3.

As explained at the end of Chapter 3, a ***regular ruled surface*** is a surface with a single coordinate patch of the form

$$\mathbf{x}(t,s) = \alpha(t) + s\mathbf{r}(t),$$

where $\alpha(t)$ is a smooth curve parametrized by arclength and, at each point of the curve, $\mathbf{r}(t)$ is a unit vector such that

1. $\mathbf{r}(t)$ is a differentiable function of t, and
2. each point $\alpha(t)$ is in the interior of an (extrinsically) straight segment in M that is parallel to $\mathbf{r}(t)$, and
3. the vectors, $\mathbf{x}_1(t,s) = \alpha'(t) + s\mathbf{r}'(t)$, $\mathbf{x}_2(t,s) = \mathbf{r}(t)$ form a basis for the tangent space.

a. *Show that a regular ruled surface is **developable*** (that is, *isometric to a region in the plane*) *if and only if*

$$[\mathbf{r}(t),\mathbf{r}'(t),\alpha'(t)] = 0, \text{ for all } t,$$

where $[\mathbf{r}(t),\mathbf{r}'(t),\alpha'(t)]$ *denotes the triple product, which by* **A.5.2** *in the Appendix A, is equal to* $\langle \mathbf{r}(t) \times \mathbf{r}'(t), \alpha'(t)\rangle$.

[Hint: Show that in this setting

$$K = 0 \Leftrightarrow \det\begin{pmatrix} \langle \mathbf{x}_{11},\mathbf{n}\rangle & \langle \mathbf{x}_{12},\mathbf{n}\rangle \\ \langle \mathbf{x}_{21},\mathbf{n}\rangle & \langle \mathbf{x}_{22},\mathbf{n}\rangle \end{pmatrix} = 0,$$

and note that (see **A.5.2**) $\langle \mathbf{V},\mathbf{n}\rangle = [\mathbf{V},\mathbf{x}_1,\mathbf{x}_2]$.]

b. *Let* $\alpha(t)$ *be a smooth curve parametrized by arclength on the surface M. If* α *has nonzero normal curvature* κ_n *at every point, then, for* $|s|$ *sufficiently small, show that*

$$\mathbf{x}(t,s) = \alpha(t) + s\frac{\mathbf{n}(\alpha(t)) \times \mathbf{n}'(\alpha(t))}{|\mathbf{n}'(\alpha(t))|}$$

is a developable regular ruled surface, that is tangent to M along α.

[Hint: Check that it is regular and then developable, and then calculate the normal vector to the ruled surface along α. You may need some of the formulas in the *Appendix* **A.5**.]

c. *Show that on a smooth surface M, if* α *is a geodesic with nonzero normal curvature* κ_n *at each point, then a ribbon can be laid flat along* α.

The properties of ruled surfaces discussed in this problem were mostly worked out in the nineteenth century, but the applications to the Ribbon Test (and the Ribbon Test, itself) are, as far as I can tell, first published in this book and were apparently not known (or, at least, not widely known) before.

PROBLEM 7.7. Curvature of the Hyperbolic Plane

What is the Gaussian curvature of the hyperbolic plane constructed out of annular strips? Calculate the Gaussian curvature in three different ways:

a. *using the extrinsic definition from Problem* **6.1** *and using the formulas from Problem* **6.2.f**.

[Hint: To use the extrinsic description you must first find a particular extrinsic embedding of a portion of the surface such as in Problem **3.1.f**.]

b. *using the intrinsic description in terms of local coordinates in Problem 7.1.*

[Hint:

$$K = -h_{22}/h, \text{ where } |\mathbf{x}_1| = h,\ |\mathbf{x}_2| = 1,\ \langle \mathbf{x}_1, \mathbf{x}_2 \rangle = 0.$$

Pick the local coordinates so that the coordinate curves $\mathbf{x}(u^1,b)$ follow the annular strips, and the coordinate curves $\mathbf{x}(a,u^2)$ are perpendicular to the annular strips.]

c. *using the intrinsic calculation from Problem* **5.7.d**.

d. *Discuss the differences between these three methods and how each affects your understanding.*

Chapter 8

Intrinsic Local Descriptions and Manifolds

In Chapter 5 we developed geometrically intrinsic descriptions of holonomy, parallel transport, and curvature of a surfaces. In Chapter 6 we developed extrinsic descriptions of Gaussian curvature and showed that it was the same as the intrinsic curvature for all C^2 surfaces. In Chapter 7, we found intrinsic local descriptions of Gaussian (intrinsic) curvature with respect to extrinsically defined local coordinates, using (extrinsic) directional derivatives. Now, in this chapter we will develop an intrinsic directional derivative that will allow intrinsic local descriptions of parallel transport. Then we will introduce the notion of manifolds that may have only intrinsically defined local coordinates. We will then put this all together to find for manifolds intrinsic local descriptions of the important intrinsic notions: covariant derivatives, geodesics, parallel transport, holonomy, Gaussian curvature, and others.

Problem 8.1. Covariant Derivative and Connection

If $\mathbf{X}_p$ is a tangent vector at the point p in M, and $\mathbf{f}$ is a vector field (a function that gives a tangent vector at each point) defined near p, then the directional derivative $\mathbf{X}_p\mathbf{f}$ is not in general a tangent vector and, thus, is not intrinsic. But we can define an intrinsic directional derivative by slightly modifying the definition of $\mathbf{X}_p\mathbf{f}$. In particular, if $\alpha(t)$ is a curve in M with $\alpha(0) = p$ and $\alpha'(0) = \mathbf{X}_p$, then

$$\mathbf{X}_p\mathbf{f} = \lim_{\delta\to 0} \frac{1}{\delta}[\mathbf{f}(\alpha(\delta)) - \mathbf{f}(p)].$$

This fails to be intrinsic only in the vector subtraction

$$[\mathbf{f}(\alpha(\delta)) - \mathbf{f}(p)].$$

Even in Euclidean space this subtraction does not literally make sense, because $\mathbf{f}(\alpha(\delta))$ is a (free) vector with base at the point $\alpha(\delta)$, and $\mathbf{f}(p)$ is a (free) vector with base at p. So in Euclidean space we perform the subtraction by first parallel translating $\mathbf{f}(\alpha(\delta))$ to a (bound) vector $\mathbf{f}(\alpha(\delta))_p$ based at p. (See Figure 8.1.) We can more correctly define

$$\mathbf{X}_p\mathbf{f} = \lim_{\delta\to 0}\frac{1}{\delta}[\mathbf{f}(\alpha(\delta))_p - \mathbf{f}(p)_p].$$

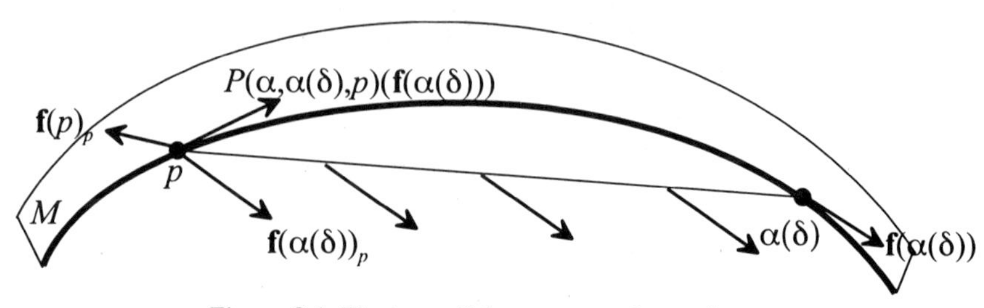

Figure 8.1. First parallel transport, then subtract.

Now this fails to be intrinsic because, even though $\mathbf{f}(\alpha(\delta))_p$ is a vector based at p, it will not in general be a vector that is tangent to M. To correct this situation we parallel transport $\mathbf{f}(\alpha(\delta))$ along α to $p = \alpha(0)$. (See Figure 8.1.) In Problem **5.4** we called this transported vector

$$P(\alpha,\alpha(\delta),p)\mathbf{f}(\alpha(\delta))$$

and it is a tangent vector at p. Now

$$[P(\alpha,\alpha(\delta),p)\mathbf{f}(\alpha(\delta)) - \mathbf{f}(p)]$$

is intrinsic since the substraction takes place in the tangent space T_pM. It is technically convenient to use, instead of that substraction,

$$[\mathbf{f}(\alpha(\delta)) - P(\alpha,p,\alpha(\delta))\mathbf{f}(p)].$$

This is allowable, because (see Problem **5.4.b**) the change in the angle between the transported vector and the velocity vector of α depends on the geodesic curvature of α and not on which vector is being transported. Thus, $P(\alpha,q,p)$ defines an isometry from the tangent plane T_qM to the tangent plane T_pM. Since the rate of change of the parallel vector field exists (and is perpendicular to the tangent plane), this isometry is continuous and

$$[P(\alpha,\alpha(\delta),p)\mathbf{f}(\alpha(\delta)) - \mathbf{f}(p)] = P(\alpha,\alpha(\delta),p)[\mathbf{f}(\alpha(\delta)) - P(\alpha,p,\alpha(\delta))\mathbf{f}(p)].$$

We then define the ***intrinsic directional derivative*** (often called the ***covariant derivative***) to be:

$$\nabla_{\mathbf{X}}\mathbf{f} = \lim_{\delta \to 0} \frac{1}{\delta}[\mathbf{f}(\alpha(\delta)) - P(\alpha, p, \alpha(\delta))\mathbf{f}(p)].$$

Let **f** *be a (tangent) vector field defined in a neighborhood of the point p on the surface M, then:*

a. *Show that if* **X** *is a tangent vector at p, then the intrinsic derivative in the direction of* **X** *is the projection of the extrinsic directional derivative onto the tangent space to the surface,* T_pM. *That is,*

$$\nabla_{\mathbf{X}}\mathbf{f} = \mathbf{Xf} - \langle \mathbf{Xf}, \mathbf{n}(p)\rangle \mathbf{n}.$$

[Hint: Use Problem **5.4**.]

In many books this is taken as the definition of the covariant derivative.

b. *If* $\gamma(s)$ *is a unit speed smooth curve with tangent vector* $\mathbf{T} = \gamma'(0)$, *then show that* $\nabla_{\mathbf{T}}\gamma'(s)$ *is the intrinsic curvature vector* κ_g *of* γ *at* $s = 0$.

[Hint: Use Part **a**.]

c. *Show that* $\mathbf{V}(s)$ *is a parallel vector field along* γ *if and only if*

$$\nabla_{\gamma'(s)}\mathbf{V} = 0, \textit{for all } s.$$

[Hint: Use Part **a** and Problem **5.4**.]

d. *Show that for fixed* **f**, *the covariant derivative*

$$\mathbf{X} \to \nabla_{\mathbf{X}}\mathbf{f}$$

is a linear operator, that is

$$\nabla_{\mathbf{X}+\mathbf{Y}}\mathbf{f} = \nabla_{\mathbf{X}}\mathbf{f} + \nabla_{\mathbf{Y}}\mathbf{f} \text{ and } \nabla_{a\mathbf{X}}\mathbf{f} = a\nabla_{\mathbf{X}}\mathbf{f}.$$

[Hint: Use Part **a**.]

e. *For a real number r and a real-valued function f, show that*

$$\nabla_{\mathbf{X}} r\mathbf{Y} = r\nabla_{\mathbf{X}}\mathbf{Y} \textit{ and } \nabla_{\mathbf{X}} f\mathbf{Y} = (\mathbf{X}f)\mathbf{Y} + f\nabla_{\mathbf{X}}\mathbf{Y}\ .$$

[Hint: Use Part **a** and Problem **4.8**.]

In many treatments of differential geometry any function ∇, for which

- ∇: $(\mathbf{X},\mathbf{Y}) \to \nabla_{\mathbf{X}}\mathbf{Y}$, where $\mathbf{X}$, $\mathbf{Y}$, $\nabla_{\mathbf{X}}\mathbf{Y}$ are vector fields on M,

and

- ∇ satisfies **8.1.d** and **8.1.e,**

is called a ***connection*** on M. There is clearly a close relationship between covariant derivatives and connections. In the literature the terms are sometimes used interchangeably, with the term *connection* used when the abstract properties in **8.1.d** and **8.1.e** are being emphasized, and the term *covariant derivative* when its role of describing rates of change is emphasized. We use the word 'connection' because a connection allows to connect the tangent vector spaces T_pM and T_qM at two different points. This is done via parallel transport (defined as we have seen in **8.1.c**) along a curve α that joins p to q. Thus any curve from p to q determines a linear transformation

$$P(\alpha,p,q)\colon T_pM \to T_qM,$$

where, for each $\mathbf{V}_p$ in T_pM, $\mathbf{V}(t) = P(\alpha,p,t)\mathbf{V}_p$ is a parallel vector field along α.

*PROBLEM 8.2. Manifolds–Intrinsic and Extrinsic

In Chapter 1, we have already seen surfaces in 3-space, $\mathbb{R}^3$, with extrinsic local coordinates, and surfaces in $\mathbb{R}^3$ with intrinsic local coordinates. We can say that extrinsic local coordinates, $\mathbf{x}\colon \mathbb{R}^2 \to \mathbb{R}^3$, are differentiable (or C^k) if $\mathbf{x}$ is differentiable (or C^k) as a function from $\mathbb{R}^2$ to $\mathbb{R}^3$. (See, for example, Problems **4.4** and **4.8**). But, for intrinsic local coordinates, no such definition is directly possible since intrinsic local coordinates are not defined in terms of a coordinate system in $\mathbb{R}^3$.

This problem can be seen clearly in the case of the annular hyperbolic plane, H. If we have a function $f\colon \mathbb{R} \to H$ (such as the parametrization of a curve on H), what would it mean to say that f is differentiable?

Locally we can (see Problem **3.1.f**) embed certain neighborhoods in H as a smooth surface in $\mathbb{R}^3$, but (as explained in **3.1.f**) it is not possible to embed the entire H as a smooth surface. And, even if it were possible to embed H as a smooth surface, its intrinsic description does not include such an embedding. So, is there any way we can say that $f: \mathbb{R} \to H$ is differentiable? There is one coordinate chart $\mathbf{x}: \mathbb{R}^2 \to H$ for all of H (see Problem **1.8.b**) that is one-to-one, and so $\mathbf{x}^{-1}$ is defined. Then $\mathbf{x}^{-1} \circ f$ is a function from $\mathbb{R}$ to $\mathbb{R}^2$ and we can ask if it is differentiable or not. This leads to a definition:

> If $\mathbf{x}: U \subset \mathbb{R}^n \to M$ is a local coordinate chart, and if $f: V \subset \mathbb{R}^m \to M$ is a function, then we say that f ***is differentiable (or*** C^k***) with respect to the chart*** $\mathbf{x}$ if $\mathbf{x}^{-1} \circ f$ is differentiable (or C^k) where it is defined (which is on $f^{-1}[\mathbf{x}(U) \cap f(V)]$).

Recall from multivariable analysis that a function f from $\mathbb{R}^m$ to $\mathbb{R}^n$ is differentiable at p if there is a linear transformation df (called the ***differential*** of f) from $\mathbb{R}^m$ to $\mathbb{R}^n$ such that, given any error (tolerance) τ there is a radius ρ such that $|x - p| < \rho$ implies

$$|f(x) - [f(p) + df(x-p)]| \leq |x - p|\tau.$$

That df is of maximal rank is equivalent to df taking $\mathbb{R}^m$ linearly onto an m-dimensional subspace of $\mathbb{R}^n$. (See Appendix B.)

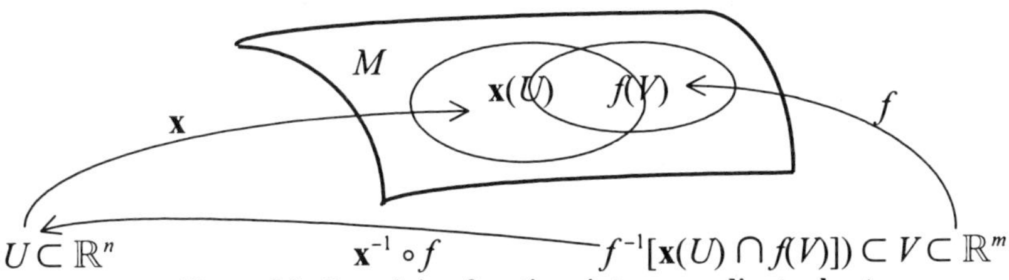

Figure 8.2. Examining functions into a coordinate chart.

We now need to check that this definition is compatible with other definitions:

***a.** *Let M be a smooth surface in $\mathbb{R}^n$ and $\mathbf{x}: \mathbb{R}^2 \to M$ be a C^1 (or C^2) local coordinate chart* (see Problems **4.4** and **4.8**, and note that the differential $d\mathbf{x}$ must have maximal rank) *with image $U = \mathbf{x}(\mathbb{R}^2)$. Then a function $f: \mathbb{R}^m \to U$ is* C^1

(or C^2*) with respect to* **x** *if and only if f is* C^1 *(or* C^2*) as a function from* $\mathbb{R}^m$ *to* $\mathbb{R}^n$*.*

Outline of a proof of **8.2.a**:

1. First prove this in the case that the chart is a Monge patch **y**. (What is the inverse of a Monge patch?) (See Problem **3.1**.)
2. Then look at $\mathbf{x}^{-1} \circ \mathbf{y}$ and argue that this is a one-to-one function from an open subset of $\mathbb{R}^2$ onto an open subset of $\mathbb{R}^2$.
3. The inverse of $\mathbf{x}^{-1} \circ \mathbf{y}$ is C^1 (or C^2), and then it follows that $\mathbf{x}^{-1} \circ \mathbf{y}$ is C^1 (or C^2). [You can use the Inverse Function Theorem (see Appendix B), but this is overkill in this case because the hard part of the Inverse Function Theorem is to prove that the function and its inverse are one-to-one and onto. In this case it is more direct to look at the differential $d(\mathbf{x}^{-1} \circ \mathbf{y})$.]

Now, if the surface M has two local coordinate charts, it is possible that a function will be differentiable with respect to one but not with respect to the other. (For example, let $\mathbb{R}^2$ be a surface with the identity as one chart and a non-differentiable one-to-one function from $\mathbb{R}^2$ to $\mathbb{R}^2$ as another chart.) To take care of this problem, we will require that two coordinate charts, **x** and **y**, for the surface M will be compatible in the sense that

$$\mathbf{x}^{-1} \circ \mathbf{y} \text{ and } \mathbf{y}^{-1} \circ \mathbf{x}$$

are both differentiable. Then any function that is differentiable with respect to **x** will also be differentiable with respect to **y**. [Be sure you see why this is true.]

We can now use this idea to expand the notion of surface, so that we can work intrinsically and in higher dimensions:

A ***differentiable*** [or C^k] ***n-manifold*** is a metric space M with a collection (called an ***atlas***) $\mathcal{A}$ such that:

- Each member of $\mathcal{A}$ is a chart $\mathbf{x}: U \subset \mathbb{R}^n \to M$ such that both U and $\mathbf{x}(U)$ are open, and **x** and $\mathbf{x}^{-1}$ are both continuous.
- Each point in M is contained in the image of at least one chart from $\mathcal{A}$.
- If $\mathbf{x}: U \subset \mathbb{R}^n \to M$ and $\mathbf{y}: V \subset \mathbb{R}^n \to M$ are two charts in $\mathcal{A}$, then

$$\mathbf{x}^{-1} \circ \mathbf{y} \text{ and } \mathbf{y}^{-1} \circ \mathbf{x}$$

are both differentiable [or C^k] where they are defined.

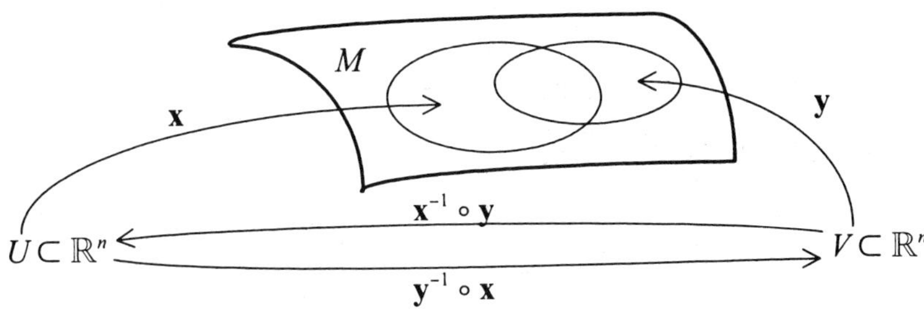

Figure 8.3. Charts in an atlas.

We say that a chart **x** for M is ***compatible*** with the atlas $\mathcal{A}$ if, for every chart **y** in $\mathcal{A}$,

$$\mathbf{x}^{-1} \circ \mathbf{y} \text{ and } \mathbf{y}^{-1} \circ \mathbf{x}$$

are both differentiable [or C^k] where they are defined.

A function $f : (M,\mathcal{A}) \to (N,\mathcal{B})$, between an m-manifold and an n-manifold are said to be [C^k] ***differentiable*** if, for every chart **x** in $\mathcal{A}$ and every chart **y** in $\mathcal{B}$,

$$\mathbf{y}^{-1} \circ f \circ \mathbf{x}$$

is [C^k] differentiable wherever it is defined as a function from a subset of $\mathbb{R}^m$ to a subset of $\mathbb{R}^n$.

Two n-manifolds, $(M,\mathcal{A})$ and $(N,\mathcal{B})$, are said to be the same, or [C^k] ***diffeomorphic***, if there is a one-to-one function $f : M \to N$, such that both f and f^{-1} are [C^k] differentiable.

If the n-manifold M is a subset of $\mathbb{R}^m$, and if each chart **x** in the atlas of M is C^k as a function from $\mathbb{R}^n$ to $\mathbb{R}^m$ and has its differential $d\mathbf{x}$ of maximal rank, then we say that M is a ***submanifold of*** $\mathbb{R}^m$.

Problem **8.2.a** shows that every smooth surface in $\mathbb{R}^m$ is a 2-manifold.

These definitions differ from those in some other books. In particular, some texts use charts that are functions from an open subset of M onto an open subset of $\mathbb{R}^n$. Such texts tend to use charts from $\mathbb{R}^n$ to the manifold for manifolds that are described extrinsically as submanifolds in a higher-dimensional Euclidean space, and to use charts from the manifold to $\mathbb{R}^n$ for manifolds that are described intrinsically. In our text, our discussion intertwines the two types of manifolds, and thus, it seems to make sense to use the same direction for the charts. In addition, some texts instead of an *atlas* require a ***maximal atlas*** for M (that is, an atlas $\mathcal{A}$ such that, if $\mathbf{x}$ is a coordinate chart for M, which is not in $\mathcal{A}$, then there is a coordinate chart $\mathbf{y}$ in $\mathcal{A}$ such that either (or both) of

$$\mathbf{x}^{-1} \circ \mathbf{y} \text{ and } \mathbf{y}^{-1} \circ \mathbf{x}$$

are both differentiable [or C^k] where they are defined). We will avoid talking about maximal atlases because atlases can be explicitly constructed, whereas maximal atlases can, in general, only be posited by using the following (easily proved) non-constructive result:

LEMMA. *If $\mathcal{A}$ is a C^k atlas for the manifold M, then the set of all charts that are compatible with $\mathcal{A}$ is the unique maximal atlas containing $\mathcal{A}$.*

We shall first look at the case of an extrinsically defined submanifold M in $\mathbb{R}^m$. Clearly the graph of a C^k function f from n-space to m-space is an (extrinsic) n-submanifold in $(n+m)$-space with an atlas consisting of the single chart $\mathbf{x}(p) = (p, f(p))$. We also can prove the following converse:

b. *If M is a C^k n-submanifold of $(n+m)$-space, and p is any point in M, then some neighborhood of p is the graph of a C^k function (that is, p has a C^k Monge patch).*

[Hint: Look at the projection π, which takes a neighborhood of p onto the tangent space at p, and apply the Inverse Function Theorem (Appendix **B.2**).]

c. *If M in $\mathbb{R}^n$ has local (extrinsic) charts that are C^k and maximal rank, then show that M is a C^k manifold with the extrinsic charts as its atlas.*

[Hint: Use the same idea of proof as in Part **a**.]

d. *Show that any surface covered by a single chart is a C^k 2-manifold, for all k. (This in particular applies to the* *annular hyperbolic plane and any open subset of Euclidean space.) Further, show that the two charts for the annular hyperbolic plane in Problem **1.8** are compatible.*

There is another class of manifolds that are defined by implicit equations. For example, the unit n-sphere is the solution of the equation

$$\sqrt{(x^1)^2 + (x^2)^2 + \cdots + (x^n)^2 + (x^{n+1})^2} = 1.$$

As a consequence of the Implicit Function Theorem in analysis, we have (see Appendix **B.3**):

THEOREM. *Let $F : \mathbb{R}^n \to \mathbb{R}^{n-m}$ be a C^1 function, and suppose $dF(x)$ has maximal rank $n - m$ at every point on a level set*

$$M = \{ x \mid F(x) = c \}.$$

Then M is a C^1 m-submanifold of $\mathbb{R}^n$.

We can now extend to (intrinsic) manifolds the intrinsic notions that we have covered in this text. The trick is merely to use the intrinsically defined forms, which we have found throughout our investigations.

For example, let us look at the notion of ***tangent vector*** and ***tangent space***. If M is an extrinsic manifold in $\mathbb{R}^n$ with C^1 chart $\mathbf{x}$ for a neighborhood of the point p, then (as in Problem **4.1**) every tangent vector at p (which is extrinsic) is the velocity vector of a curve in M. And it is easy to see that, if γ and λ are two C^1 curves in M with $\gamma(0) = \lambda(0) = p$, then they have the same velocity vectors, $\gamma'(0) = \lambda'(0)$, if and only if $\mathbf{x}^{-1} \circ \gamma$ and $\mathbf{x}^{-1} \circ \lambda$ have the same velocity vectors, $(\mathbf{x}^{-1} \circ \gamma)'(0) = (\mathbf{x}^{-1} \circ \lambda)'(0)$, as curves in $\mathbb{R}^n$. Thus we can define the ***(intrinsic) tangent space***, $T_p M$, at p in M to be the equivalence classes of C^1 curves γ in M with $\gamma(0) = p$ with the equivalence relation

$$\gamma \approx \lambda \text{ if and only } (\mathbf{x}^{-1} \circ \gamma)'(0) = (\mathbf{x}^{-1} \circ \lambda)'(0).$$

Each equivalence class of curves in the tangent space is then called an ***(intrinsic) tangent vector***, and we will denote the equivalence class of γ by the notation $[\gamma]$.

e. *If the manifold M has an atlas, show that the definition of the intrinsic tangent space at p does not depend on which* *chart (containing p) you choose from the atlas. Show also that for each chart* **y** *(containing p), the function from the tangent space of* $\mathbb{R}^n$ *at* $q = \mathbf{y}^{-1}(p)$ *to* T_pM *defined by*

$$d\mathbf{y}(\mathbf{X}_q) = [t \rightarrow \mathbf{y}(q + t\mathbf{X}_q)]$$

is one-to-one and onto. Use this to define a vector space structure on T_pM *that is independent of which chart (containing p) you use.*

Now, how do we visualize this intrinsically? The answer is that we do it naturally all the time! Our three-dimensional physical universe is a 3-manifold, which according to physicists is not Euclidean three space. However, in the very small neighborhood of 3-space in which we physically move our bodies and draw pictures, we have no problems drawing the usual pictures of vectors, and for curves (in our normal physical experience), we have no trouble imagining a tangent vector as a straight line (geodesic) segment with an arrowhead on one end. This makes sense because the space near us is indistinguishable from a region in Euclidean 3-space.

If we assume that each point p in n-manifold M has a neighborhood that intrinsically is indistinguishable from a region in n-space, then we can use it to define a ***Riemannian metric*** as in Problem **4.3**—this is the case with the annular hyperbolic plane. Or we can posit a symmetric, bilinear, positive definite, real-valued function $\langle \mathbf{X},\mathbf{Y} \rangle_p$ that varies C^k with respect to p and call it a Riemannian metric. Then we can use Riemannian metric to define angles and lengths by setting

$$\langle \mathbf{X},\mathbf{Y} \rangle = |\mathbf{X}|\,|\mathbf{Y}| \cos \theta_{\mathbf{XY}}.$$

As always, we consider a function defined on M or, in this case, on the tangent vectors of M to be C^k if, using a chart **y**, the corresponding functions on $\mathbb{R}^n$ (or in this case its tangent vectors) are C^k. In fact, we can consider the collection of all tangent vectors on M to be a differentiable manifold (of dimension n^2) called the ***tangent bundle*** TM ***of*** M. The tangent bundle $T\mathbb{R}^n$ of $\mathbb{R}^n$ is the collection of all *bound* vectors on $\mathbb{R}^n$. Since at each point q in $\mathbb{R}^n$ the vectors bound at q form an n-dimensional space and, thus, the tangent bundle has n^2 dimensions. Then the map $d\mathbf{x}$ (defined in Part **e**) maps a vector $\mathbf{X}_q$ bound at q to the tangent vector

$d\mathbf{x}(\mathbf{X}_q)$ in $T_{\mathbf{x}(q)}$. Then the atlas for TM consists of a chart $d\mathbf{x}$ for each chart $\mathbf{x}$ in the atlas for M.

Directional derivatives of real-valued functions on an n-manifold M are intrinsic because they are just real numbers, and the definition (after Problem **4.5**)

$$\mathbf{X}_p f = \frac{d}{dt} f(\gamma(t))_{t=0} = \lim_{h \to 0} \frac{f(\gamma(h)) - f(\gamma(0))}{h}$$

works unchanged where $\mathbf{X}_p = [\gamma]$.

Directional derivatives of vector valued functions (vector fields) are not intrinsic because $\mathbf{X}_p\mathbf{Y}$ will not in general be a tangent vector, so there is no hope of defining them on an intrinsic manifold; however, we can define the (***intrinsic***) ***covariant derivative*** on any manifold M in the following ways:

> If M is an (extrinsic) m-dimensional manifold in n-space, and if $\mathbf{X}$ is a (tangent) vector in the tangent space T_pM at the point p in M, and if $\mathbf{f}$ is a tangent vector *field*, then the ***covariant derivative*** of $\mathbf{f}$ with respect to $\mathbf{X}$ can be extrinsically defined as
>
> $$\nabla_{\mathbf{X}}\mathbf{f} = \{\text{projection of } \mathbf{Xf} \text{ onto the tangent space } T_pM\}.$$
>
> If M is an intrinsic m-manifold, and we have an intrinsic notion of parallel transport (as with the annular hyperbolic plane), then we can define the ***covariant derivative*** as in Problem **8.1**. Or, we can define a ***connection*** on M as any function
>
> $$\nabla: (\mathbf{X},\mathbf{Y}) \to \nabla_{\mathbf{X}}\mathbf{Y}, \text{ where } \mathbf{X}, \mathbf{Y}, \nabla_{\mathbf{X}}\mathbf{Y} \text{ are vector fields on } M,$$
>
> which satisfies **8.1.d** and **8.1.e**.

None of these ways allows us to intrinsically compute the covariant derivative from the knowledge of a coordinate chart and a Riemannian metric—this is the deficiency we will correct in the next problem.

PROBLEM 8.3. *Christoffel Symbols*

If $\mathbf{x}$ is a local coordinate system for a neighborhood of p in the manifold M, then the covariant derivative $\nabla_{\mathbf{x}_i}\mathbf{x}_j$ is a tangent vector to M and thus is a linear combination of the basis vectors $\mathbf{x}_1, \mathbf{x}_2, \ldots, \mathbf{x}_m$:

$$\nabla_{\mathbf{x}_i}\mathbf{x}_j = \Gamma_{ij}^1\mathbf{x}_1 + \Gamma_{ij}^2\mathbf{x}_2 + \cdots + \Gamma_{ij}^m\mathbf{x}_m = \sum_k \Gamma_{ij}^k\mathbf{x}_k.$$

The coefficients Γ_{ij}^k are called the ***Christoffel symbols*** and are clearly intrinsic quantities. In this problem we wish to find intrinsic formulae of these symbols and therefore of the intrinsic derivative, but along the way we will need to use the extrinsic description in Problem **8.1.a**. So, until we give a purely intrinsic description at the end of this problem, the manifold M is extrinsic.

a. *Show that*

$$\Gamma_{ij}^k = \sum_l \langle \mathbf{x}_{ij}, \mathbf{x}_l \rangle g^{lk},$$

where the matrix (g^{lk}) *is the inverse of the matrix* (g_{lk}). Many texts take this latter expression as the definition of the Christoffel symbols, but I believe such a definition hides the geometric meaning.

Outline of solution:

1. Argue that

$$\langle \mathbf{x}_{ij}, \mathbf{x}_l \rangle = \langle \nabla_{\mathbf{x}_i}\mathbf{x}_j, \mathbf{x}_l \rangle = \sum_k \Gamma_{ij}^k g_{kl}.$$

2. Using the fact that the matrix (g^{lk}) is the inverse of the matrix (g_{lk}), show that

$$\sum_l \langle \mathbf{x}_{ij}, \mathbf{x}_l \rangle g^{lm} = \sum_l \left(\sum_k \Gamma_{ij}^k g_{kl} \right) g^{lm} = \Gamma_{ij}^m.$$

b. *Explain each step of the following argument:*
If $\mathbf{Y} = \sum Y^j\mathbf{x}_j$ *is a (tangent) vector field (note that the* Y^j *are real valued functions), then*

$$\nabla_{\mathbf{x}_i}\mathbf{Y} = \sum_j \nabla_{\mathbf{x}_i}(Y^j\mathbf{x}_j) = \sum_j [(\mathbf{x}_i Y^j)\mathbf{x}_j + Y^j(\nabla_{\mathbf{x}_i}\mathbf{x}_j)] =$$

$$= \sum_j \left[(\mathbf{x}_i Y^j)\mathbf{x}_j + Y^j \left(\sum_k \Gamma_{ij}^k \mathbf{x}_k \right) \right] = \sum_k \left(\mathbf{x}_i Y^k + \sum_j \Gamma_{ij}^k Y^j \right) \mathbf{x}_k.$$

Some texts use

$$\nabla_{\mathbf{x}_i}\mathbf{Y} = \sum_k \left(\mathbf{x}_i Y^k + \sum_j \Gamma_{ij}^k Y^j \right) \mathbf{x}_k$$

as the definition of the covariant derivative.

c. *Show that*

$$\langle \mathbf{x}_{ij}, \mathbf{x}_k \rangle = \tfrac{1}{2}[\mathbf{x}_i g_{jk} - \mathbf{x}_k g_{ji} + \mathbf{x}_j g_{ki}]$$

and thus

$$\Gamma_{ij}^k = \tfrac{1}{2}\sum_l g^{kl}[\mathbf{x}_j g_{il} - \mathbf{x}_l g_{ij} + \mathbf{x}_i g_{lj}].$$

[Hint:

$$\begin{aligned}\langle \mathbf{x}_{ij}, \mathbf{x}_k \rangle &= \mathbf{x}_i\langle \mathbf{x}_j, \mathbf{x}_k \rangle - \langle \mathbf{x}_j, \mathbf{x}_{ik} \rangle = \\ &= \mathbf{x}_i\langle \mathbf{x}_j, \mathbf{x}_k \rangle - (\mathbf{x}_k\langle \mathbf{x}_j, \mathbf{x}_i \rangle - \langle \mathbf{x}_{kj}, \mathbf{x}_i \rangle) = \\ &= \mathbf{x}_i\langle \mathbf{x}_j, \mathbf{x}_k \rangle - \mathbf{x}_k\langle \mathbf{x}_j, \mathbf{x}_i \rangle + \mathbf{x}_j\langle \mathbf{x}_k, \mathbf{x}_i \rangle - \langle \mathbf{x}_k, \mathbf{x}_{ji} \rangle .]\end{aligned}$$

We started out with an extrinsic definition of the covariant derivative and thus the Christoffel symbols. But now we have an intrinsic description of the Christoffel symbols, and so we can give:

Intrinsic Definition of the covariant derivative. *Let* **x** *be a* C^2 *local coordinate chart for the open set* U *in the* n*-manifold* M. *If*

$$\mathbf{X}_p = \sum_i X^i \mathbf{x}_i(p)$$

is a tangent vector at p, *and*

$$\mathbf{f}(p) = \sum_j F^j(p)\mathbf{x}_j(p)$$

is a differentiable function defined on U, *then*

$$\nabla_{\mathbf{X}_p}\mathbf{f} = \sum_i X^i \nabla_{\mathbf{x}_i(p)}(\sum_j F^j \mathbf{x}_j) = \sum_i X^i(\sum_j \nabla_{\mathbf{x}_i(p)}(F^j \mathbf{x}_j)) =$$
$$= \sum_i \left[X^i \sum_k \left[\mathbf{x}_i(F^k) + \Sigma_j \left(F^j \tfrac{1}{2}\sum_l g^{kl}[\mathbf{x}_j g_{il} - \mathbf{x}_l g_{ij} + \mathbf{x}_i g_{lj}] \right) \right] \mathbf{x}_k \right].$$

Note that this definition will work on any n-manifold with a Riemannian metric. The resulting connection

∇: $(\mathbf{X},\mathbf{Y}) \to \nabla_{\mathbf{X}}\mathbf{Y}$, where $\mathbf{X}$, $\mathbf{Y}$, $\nabla_{\mathbf{X}}\mathbf{Y}$ are vector fields on M, is often called the ***Riemannian connection***. It is a theorem (see, for example, Theorem 8.6 on page 236 of [**DG:** Millman and Parker]) that this connection is the only connection on M which satisfies:

- ***Metric connection***—Parallel transport with respect to ∇ is an isometry, and

- ***Symmetric*** (or ***Torsion-free***) ***connection***—For all vector fields $\mathbf{X}$ and $\mathbf{Y}$ on M and for all real-valued functions f on M, we have

$$(\nabla_{\mathbf{X}}\mathbf{Y} - \nabla_{\mathbf{Y}}\mathbf{X})f = \mathbf{X}_p(\mathbf{Y}f) - \mathbf{Y}_p(\mathbf{X}f).$$

(See Problem **8.5.c**.)

e. *Compute for geodesic rectangular (or polar) coordinates on any surface that*

$$\Gamma_{11}^{1} = \tfrac{1}{2}g^{11}[\mathbf{x}_1 g_{11} - \mathbf{x}_1 g_{11} + \mathbf{x}_1 g_{11}] = \tfrac{1}{2}h^{-2}\mathbf{x}_1(h^2) = h_1/h,$$
$$\Gamma_{11}^{2} = -hh_2,\ \Gamma_{12}^{1} = \Gamma_{21}^{1} = h_2/h,$$
$$\textit{all others zero.}$$

Evaluate in the special case of a sphere.

Problem 8.4. Intrinsic Curvature and Geodesics

Now we will use our description of covariant derivatives in terms of local coordinates to find intrinsic local coordinate descriptions of the geodesic (intrinsic) curvature of a curve and thus of geodesics.

a. *If $\gamma(s)$ is a curve parametrized by arclength, then, according to Problem* **8.1.b**, *the intrinsic curvature at $\gamma(a)$ is given by*

$$\kappa_g(a) = \nabla_{\gamma'(a)}\gamma'.$$

Show that if you express the curve in terms of a local coordinates $\mathbf{x}$ *as*

$$\gamma(s) = \mathbf{x}(\gamma^1(s), \gamma^2(s)),$$

then

$$\gamma'(s) = (\gamma^1)'_s \mathbf{x}_1 + (\gamma^2)'_s \mathbf{x}_2,$$

and the intrinsic curvature is given by

$$\kappa_g(a) = \sum_k \left[(\gamma^k)''_a + \sum_{i,j} \Gamma^k_{ij}(\gamma(a))(\gamma^i)'_a(\gamma^j)'_a \right] \mathbf{x}_k.$$

[Hint: Use the fact that, for any real-valued function $f(s)$,

$$\nabla_{\gamma'(a)} f(s)|_{s=a} = f'(a).]$$

b. *Show that if $\gamma(s)$ is a curve parametrized by arclength then, γ is a geodesic if and only if*

$$(\gamma^k)''_s + \sum_{i,j} \Gamma^k_{ij}(\gamma(s))(\gamma^i)'_s(\gamma^j)'_s = 0,$$

for each k and each point along γ.

These are the differential equations for a geodesic expressed in local coordinates. This has theoretical importance in analytic treatments of geodesics, but in practice these equations can rarely be solved except approximately.

c. *Express the results in Parts* **a** *and* **b** *in the case that* **x** *is geodesic rectangular (or polar) coordinates.*

PROBLEM 8.5. Lie Brackets, Coordinate Vector Fields

We now want to find intrinsic expressions in local coordinates for the curvature of a manifold, but first we must examine the ways in which two tangent vector fields interact.

a. *Let* **x** *be a local chart for the open set U in the* C^2 *manifold M. Show that*

$$\nabla_{\mathbf{x}_i}\mathbf{x}_j = \nabla_{\mathbf{x}_j}\mathbf{x}_i.$$

However, this commutativity does not hold in general. In fact:

b. *In* $\mathbb{R}^2$ *find two (simple) vector fields*

$$\mathbf{A}(x,y) = \mathbf{e}_1 + a(x,y)\mathbf{e}_2 \text{ and } \mathbf{B}(x,y) = \mathbf{e}_2$$

such that

$$\nabla_{\mathbf{A}(0,0)}\mathbf{B} = \mathbf{A}(0,0)\mathbf{B} \neq \mathbf{B}(0,0)\mathbf{A} = \nabla_{\mathbf{B}(0,0)}\mathbf{A}.$$

If **V** and **W** are two vector fields defined on a neighborhood of p in M, then we define the ***Lie bracket*** $[\mathbf{V},\mathbf{W}]$ by setting

$$[\mathbf{V},\mathbf{W}]_p \equiv \nabla_{\mathbf{V}(p)}\mathbf{W} - \nabla_{\mathbf{W}(p)}\mathbf{V}.$$

From **8.5.a** we know that if **V** and **W** are the coordinate vector fields of some coordinate chart, then $[\mathbf{V},\mathbf{W}]\equiv 0$. In part **d** we show that these are the only examples.

c. *Show that, even though* $\mathbf{X}_p\mathbf{Y}$ *is not a tangent vector in general,*

$$\mathbf{X}_p\mathbf{Y} - \mathbf{Y}_p\mathbf{X}$$

is a tangent vector and is equal to $[\mathbf{X},\mathbf{Y}]_p$. This is often the definition of the Lie bracket.

[Hint: Express in terms of local coordinates and use **8.1** and linearity.]

***d.** *On an n-manifold M, show that the n vector fields* $\{\mathbf{V}_i\}$ *are equal to* $\{\mathbf{x}_i\}$ *for some coordinate chart* **x** *if and only if*

$$[\mathbf{V}_j,\mathbf{V}_k] \equiv \nabla_{\mathbf{V}_j}\mathbf{V}_k - \nabla_{\mathbf{V}_k}\mathbf{V}_j = 0, \textit{ for all } 1\leq i,j\leq n.$$

Outline of a proof:

This outline assumes that the reader has a familiarity with flows defined by vector fields and with the theorem from analysis that a C^1 vector field always has a unique flow. For a discussion of these results, consult [**An**: Strichartz], Chapter 11, or [**DG**: Dodson/Poston], VII.6 and VII.7. In the latter, the details of this outline are filled in.

1. Given a C^1 vector field **V** defined and nonzero in a neighborhood of p in M, then there is a coordinate chart **x** such that $\mathbf{V} = \mathbf{x}_1$.
2. If **V** and **W** are two C^1 vector fields on M with flows ϕ_s and ψ_s then the flows commute

$$\phi_a \circ \psi_b = \psi_b \circ \phi_a, \text{ wherever defined}$$

if and only if

$$[\mathbf{V},\mathbf{W}]_p = \mathbf{0}, \text{ for all } p.$$

3. Use the flows to define the coordinate chart $\mathbf{x}$.

PROBLEM 8.6. Riemann Curvature Tensors

We now want to extend the notion of Gaussian (intrinsic) curvature to n-manifolds. First we express the Gaussian curvature of a surface in terms of the covariant derivative in local coordinates. For this problem we assume that $\mathbf{x}$ is a local *orthogonal* C^2 coordinate system on M.

In spite of **8.5.a**, in general,

$$\nabla_{\mathbf{x}_1}\nabla_{\mathbf{x}_2}\mathbf{V} \neq \nabla_{\mathbf{x}_2}\nabla_{\mathbf{x}_1}\mathbf{V}.$$

In fact, you can prove the following result.

a. *On a surface M with orthogonal coordinates* $\mathbf{x}(u^1,u^2)$, *let* $\mathbf{V}$ *be a tangent vector field such that* $\nabla_{\mathbf{x}_2}\nabla_{\mathbf{x}_1}\mathbf{V}$ *and* $\nabla_{\mathbf{x}_1}\nabla_{\mathbf{x}_2}\mathbf{V}$ *exist and are continuous. Then, at every p in M, show that*

$$|\nabla_{\mathbf{x}_1}\nabla_{\mathbf{x}_2}\mathbf{V} - \nabla_{\mathbf{x}_2}\nabla_{\mathbf{x}_1}\mathbf{V}| = |\mathbf{V}||\mathbf{x}_1||\mathbf{x}_2||K(p)|,$$

where K(p) is the Gaussian curvature!

Outline of a proof:

1. Let $p = \mathbf{x}(0,0)$. Since the covariant derivative and the intrinsic curvature can both be defined in terms of parallel transport, look at parallel transport along the coordinate curves and use the following abbreviations:

$$P_1(\delta,a) = P(t \to x(t,a), \mathbf{x}(0,a), \mathbf{x}(\delta,a)),$$
$$P_2(a,\delta) = P(t \to x(a,t), x(a,0), x(a,\delta)).$$

Look at the situation in Figure 8.4 and define

$$\mathbf{P}(\varepsilon,\delta) = P_1(\varepsilon,\delta)[P_2(0,\delta)\mathbf{V}(p)] - P_2(\varepsilon,\delta)[P_1(\varepsilon,0)\mathbf{V}(p)].$$

Then starting with

$$K(p) = \lim_{R\to 0}(\mathscr{H}(R)/A(R)) = \lim_{R\to 0}(\theta/A(R)),$$

and using the definition of area as an integral (Problem **4.5**), show that

$$|\mathbf{V}||\mathbf{x}_1||\mathbf{x}_2||K(p)| = \lim_{\varepsilon,\delta\to 0} \frac{|\mathbf{P}(\varepsilon,\delta)|}{\varepsilon\delta}.$$

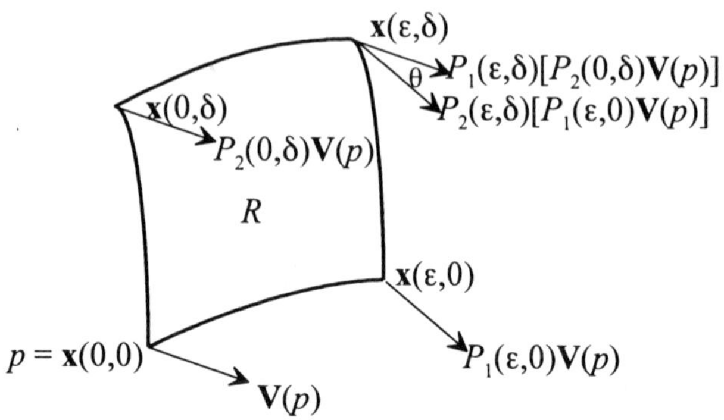

Figure 8.4. Parallel transport along coordinate curves.

2. Now, denoting $\mathbf{V}(\mathbf{x}(a,b)) = \mathbf{V}(a,b)$, use the limit definition of covariant derivative, the fact that $\nabla_{\mathbf{x}_1}$ is continuous, and the fact that parallel transport is a linear isometry to compute

$$\nabla_{\mathbf{x}_1}\nabla_{\mathbf{x}_2}\mathbf{V} - \lim_{\varepsilon,\delta\to 0}\frac{\mathbf{P}(\varepsilon,\delta)}{\varepsilon\delta} =$$
$$= \lim_{\delta\to 0}\nabla_{\mathbf{x}_1}\left(\tfrac{1}{\delta}[\mathbf{V}(a,\delta) - P_2(a,\delta)\mathbf{V}(a,0)]\right) - \lim_{\varepsilon,\delta\to 0}\frac{\mathbf{P}(\varepsilon,\delta)}{\varepsilon\delta}.$$

Expand this expression and rearrange until you get it equal to

$$\lim_{\delta\to 0}\frac{1}{\delta}\{(\nabla_{\mathbf{x}_1}\mathbf{V})(0,\delta) - P_2(\varepsilon,\delta)[(\nabla_{\mathbf{x}_1}\mathbf{V})(0,0)]\} = \nabla_{\mathbf{x}_2}((\nabla_{\mathbf{x}_1}\mathbf{V})(0,0)).$$

Then conclude the result.

b. *In part* **a** *we can set* **V** *equal to* $\mathbf{x}_1$ *and, then after parallel transport around the region, the vector* $\mathbf{x}_1$ *will change in the* $\mathbf{x}_2$ *direction (Why?) and thus (Why?)*

$$\langle(\nabla_{\mathbf{x}_1}\nabla_{\mathbf{x}_2}\mathbf{x}_1 - \nabla_{\mathbf{x}_2}\nabla_{\mathbf{x}_1}\mathbf{x}_1), \mathbf{x}_2\rangle_p = \langle\mathbf{x}_1,\mathbf{x}_1\rangle\langle\mathbf{x}_2,\mathbf{x}_2\rangle K(p).$$

[Hint: Look in Figure 8.4 at the effect of the parallel transport.]

Now we can find an expression for the Gaussian curvature of the surfaces in an n-manifold that are determined by two coordinates:

Intrinsic Definition of Sectional Curvature. *Let* $\mathbf{x}$ *be a* C^2 *local coordinate chart for the open set U in the n-manifold M. Let*

$$p = \mathbf{x}(a^1, a^2, \ldots, a^n)$$

be any point in U. For each $1 \le i,j \le n$, *the* ***sectional curvature*** *of M at p in the section determined by* $\mathbf{x}_i$ *and* $\mathbf{x}_j$ *is the Gaussian curvature of the surface,* S^{ij}, *in U containing p with local coordinates defined by*

$$S^{ij} = \mathbf{x}(a^1, \ldots, a^{i-1}, x^i, a^{i+1}, \ldots, a^{j-1}, x^j, a^{j+1}).$$

By **8.6.b**, *this sectional curvature is*

$$K_p(\mathbf{x}_i \wedge \mathbf{x}_j) = \frac{1}{|\mathbf{x}_i|^2}\frac{1}{|\mathbf{x}_j|^2}\langle(\nabla_{\mathbf{x}_i}\nabla_{\mathbf{x}_j}\mathbf{x}_i - \nabla_{\mathbf{x}_j}\nabla_{\mathbf{x}_i}\mathbf{x}_i), \mathbf{x}_j\rangle_p .$$

Thus, if $\mathbf{X}$ and $\mathbf{Y}$ are two orthogonal unit tangent vectors in an n-manifold, it makes sense to try to use the expression

$$\langle \nabla_{\mathbf{X}}\nabla_{\mathbf{Y}}\mathbf{X} - \nabla_{\mathbf{Y}}\nabla_{\mathbf{X}}\mathbf{X}, \mathbf{Y}\rangle_p$$

as the measure of curvature of the two-dimensional section determined by two tangent vectors at p, $\mathbf{X}$ and $\mathbf{Y}$. This almost works except for the problem that the expression only makes sense if $\mathbf{X}$ and $\mathbf{Y}$ are vector *fields* (otherwise you could not differentiate). You could extend $\mathbf{X}$ and $\mathbf{Y}$ to vector fields in a neighborhood of p, but unfortunately the result would depend on which vector field you choose. There is a way out this dilemma: Someone discovered (I do not know how or who) the following, which you can prove.

c. *For vector fields,* $\mathbf{X}$, $\mathbf{Y}$, *and* $\mathbf{Z}$, *the expression*

$$\mathbf{R}_p(\mathbf{Y},\mathbf{X})\mathbf{Z} \equiv \nabla_{\mathbf{X}_p}\nabla_{\mathbf{Y}}\mathbf{Z} - \nabla_{\mathbf{Y}_p}\nabla_{\mathbf{X}}\mathbf{Z} - \nabla_{[\mathbf{X},\mathbf{Y}]_p}\mathbf{Z}$$

only depends on the vectors $\mathbf{X}_p$, $\mathbf{Y}_p$, $\mathbf{Z}_p$ *and not on the rest of the fields. (Note that if* $\mathbf{X}$ *and* $\mathbf{Y}$ *are coordinate vector fields then, by* **8.5.a**, $[\mathbf{X},\mathbf{Y}] = 0$.) WARNING: Some books define $\mathbf{R}(\mathbf{Y},\mathbf{X})\mathbf{Z}$ as the negative of our definition.

[Hint: If $\mathbf{F}(\mathbf{X})$ is a vector field that depends linearly on another vector field $\mathbf{X}$, then there is a trick that works to check whether $\mathbf{F}_p(\mathbf{X})$ depends only on $\mathbf{X}_p$. Let k be any real-valued function defined in a neighborhood of p such that if $k(p) = 1$, then $\mathbf{F}_p(\mathbf{X})$ depends only on $\mathbf{X}_p$ if and only if

$$\mathbf{F}_p(k\mathbf{X}) = k(p)\mathbf{F}_p(\mathbf{X}) = \mathbf{F}_p(\mathbf{X}).$$

Note that, in this case, and because $\mathbf{F}$ is linear,

$$\text{if } \mathbf{X} = \Sigma X^i \mathbf{x}_i \text{ then } \mathbf{F}_p(\mathbf{X}) = \Sigma X^i(p)\mathbf{F}_p(\mathbf{x}_i).]$$

The function $\mathbf{R}(\mathbf{X},\mathbf{Y})\mathbf{Z}$ *is called the* ***Riemann curvature tensor field of type* (1,3)**.

It is called a *tensor of type* (1,3) because it depends linearly on its three variables and because $\mathbf{R}_p(\mathbf{X},\mathbf{Y})\mathbf{Z}$ is a vector (thus type (1,–)) that depends only on the three (thus type (–,3)) vectors $\mathbf{X}_p,\mathbf{Y}_p,\mathbf{Z}_p$. (See *Appendix* **A.8**.)

The function

$$\mathbf{R}(\mathbf{X},\mathbf{Y},\mathbf{Z},\mathbf{W}) = \langle \mathbf{R}(\mathbf{X},\mathbf{Y})\mathbf{Z},\mathbf{W}\rangle$$

is called the ***Riemann curvature tensor field of type* (0,4)**.

If $\mathbf{X}$ *and* $\mathbf{Y}$ *are two orthogonal unit tangent vectors at p in an n-manifold, then*

$$K_p(\mathbf{X} \wedge \mathbf{Y}) = \mathbf{R}_p(\mathbf{X},\mathbf{Y},\mathbf{X},\mathbf{Y})$$

is called the ***sectional curvature***.

The sectional curvature is NOT linear in $\mathbf{X}$ and $\mathbf{Y}$. In fact,

$$K_p((\mathbf{A}+\mathbf{B}) \wedge \mathbf{Y}) = \mathbf{R}_p((\mathbf{A}+\mathbf{B}),\mathbf{Y},(\mathbf{A}+\mathbf{B}),\mathbf{Y}) =$$

$$= \mathbf{R}_p(\mathbf{A},\mathbf{Y},\mathbf{A},\mathbf{Y}) + \mathbf{R}_p(\mathbf{A},\mathbf{Y},\mathbf{B},\mathbf{Y}) + \mathbf{R}_p(\mathbf{B},\mathbf{Y},\mathbf{A},\mathbf{Y}) + \mathbf{R}_p(\mathbf{B},\mathbf{Y},\mathbf{B},\mathbf{Y}),$$

which is a phenomenon that we encountered with normal curvature and the second fundamental form. This is the reason that we have to look at

Riemann tensors if we want to express sectional curvature with respect to different local coordinates.

> *If we have local coordinates* **x**, *then the Riemann curvature tensor of type* (1,3) *is determined by the* n^4 *numbers*
>
> $$R^l_{ijk}(p) = l^{\text{th}} \text{ coordinate of } \mathbf{R}_p(\mathbf{x}_i, \mathbf{x}_j)\mathbf{x}_k = \\ = l^{\text{th}} \text{ coordinate of } \nabla_{\mathbf{x}_j(p)}\nabla_{\mathbf{x}_i(p)}\mathbf{x}_k - \nabla_{\mathbf{x}_i(p)}\nabla_{\mathbf{x}_j(p)}\mathbf{x}_k.$$
>
> *or*
>
> $$\mathbf{R}(\mathbf{x}_i, \mathbf{x}_j)\mathbf{x}_k = \sum_l R^l_{ijk}\mathbf{x}_l.$$

(Remember that $[\mathbf{x}_i,\mathbf{x}_j] = 0$.)

d. *Show that the* ***Riemann curvature tensor of type* (0,4)** *is determined by the* n^4 *numbers*

$$R_{ijkh} \equiv \mathbf{R}(\mathbf{x}_i, \mathbf{x}_j, \mathbf{x}_k, \mathbf{x}_h) = \sum_l R^l_{ijk}g_{lh}.$$

The ***sectional curvature*** *of the two-dimensional subspace of* T_pM *spanned by the* ***orthonormal*** *vectors*

$$\mathbf{X} = \Sigma X^i\mathbf{x}_i \text{ and } \mathbf{Y} = \Sigma Y^j\mathbf{x}_j$$

is given by

$$K(\mathbf{X} \wedge \mathbf{Y}) = \sum_i\sum_j\sum_k\sum_h R_{ijkh}X^iY^jX^kY^h.$$

Calculation of Curvature Tensors in Local Coordinates

We will now find the coefficients of the Riemann curvature tensor with respect to local coordinates **x**:

$$\mathbf{R}(\mathbf{x}_i\mathbf{x}_j)\mathbf{x}_k = \sum_l R^l_{ijk}\mathbf{x}_l = \nabla_{\mathbf{x}_j}\nabla_{\mathbf{x}_i}\mathbf{x}_k - \nabla_{\mathbf{x}_i}\nabla_{\mathbf{x}_j}\mathbf{x}_k =$$

$$= \nabla_{\mathbf{x}_j}\left(\sum_l \Gamma^l_{ik}\mathbf{x}_l\right) - \nabla_{\mathbf{x}_i}\left(\sum_l \Gamma^l_{jk}\mathbf{x}_l\right) =$$

$$= \sum_l [(\mathbf{x}_j\Gamma^l_{ik})\mathbf{x}_l + \Gamma^l_{ik}(\nabla_{\mathbf{x}_j}\mathbf{x}_l)] - \sum_l [(\mathbf{x}_i\Gamma^l_{jk})\mathbf{x}_l + \Gamma^l_{jk}(\nabla_{\mathbf{x}_i}\mathbf{x}_l)] =$$

$$= \sum_l \left[(\mathbf{x}_j\Gamma^l_{ik})\mathbf{x}_l + \Gamma^l_{ik}\left(\sum_h \Gamma^h_{jl}\mathbf{x}_h\right)\right] - \sum_l \left[(\mathbf{x}_i\Gamma^l_{jk})\mathbf{x}_l + \Gamma^l_{jk}\left(\sum_h \Gamma^h_{il}\mathbf{x}_h\right)\right] =$$

$$= \sum_l (\mathbf{x}_j\Gamma^l_{ik})\mathbf{x}_l + \sum_l\sum_h \Gamma^l_{ik}\Gamma^h_{jl}\mathbf{x}_h - \sum_l (\mathbf{x}_i\Gamma^l_{jk})\mathbf{x}_l - \sum_l\sum_h \Gamma^l_{jk}\Gamma^h_{il}\mathbf{x}_h =$$

$$= \sum_h \left| \mathbf{x}_j\Gamma^h_{ik} + \sum_l \Gamma^l_{ik}\Gamma^h_{jl} - \mathbf{x}_i\Gamma^h_{jk} - \sum_l \Gamma^l_{jk}\Gamma^h_{il} \right| \mathbf{x}_h .$$

Therefore,

$$R^h_{ijk} = \mathbf{x}_j\Gamma^h_{ik} + \Sigma_l\, \Gamma^l_{ik}\Gamma^h_{jl} - \mathbf{x}_i\Gamma^h_{jk} - \Sigma_l\, \Gamma^l_{jk}\Gamma^h_{il} .$$

For geodesic rectangular coordinates you can calculate (using **8.3.e**):

$$R^2_{212} = 0 \text{ and } R^1_{212} = K,$$

and in agreement with **8.6.a**,

$$|\mathbf{R}(\mathbf{x}_2\mathbf{x}_1)\mathbf{x}_2| = |\mathbf{x}_1|K.$$

A similar calculation will show that

$$R^1_{121} = 0 \text{ and } R^2_{121} = h^2K.$$

For any orthogonal local coordinates, the ***Riemann Curvature Tensor of Type (4,0)*** is defined by the equation

$$R_{ijkl} = g_{ii}R^i_{jkl} .$$

Thus, for geodesic rectangular coordinates

$$R_{1212} = R_{2121} = h^2K.$$

If $\mathbf{X} = \mathbf{x}_i/|\mathbf{x}_i|$ and $\mathbf{Y} = \mathbf{x}_j/|\mathbf{x}_j|$ are unit vectors in the coordinate directions, then the ***sectional curvature*** of M is calculated as

$$K\left(\frac{\mathbf{x}_i}{|\mathbf{x}_i|} \wedge \frac{\mathbf{x}_j}{|\mathbf{x}_j|}\right) = R_{ijij}\frac{1}{|\mathbf{x}_i|^2}\frac{1}{|\mathbf{x}_j|^2} = K.$$

In particular, for geodesic rectangular coordinates, the sectional curvature (with respect to orthogonal unit vectors) is the same as the Gaussian curvature.

PROBLEM 8.7. *Intrinsic Calculations in Examples*

Find the Riemannian metric, the Christoffel symbols, the Riemann curvature tensors, and the sectional (Gaussian) curvature for:

a. *the cylinder.*

b. *the sphere.*

c. *the torus* ($\mathbf{S}^1 \times \mathbf{S}^1$) *in* $\mathbb{R}^4$ *with coordinates*

$$\mathbf{x}(u^1,u^2) = (\cos u^1, \sin u^1, \cos u^2, \sin u^2).$$

This is usually called the ***flat torus****. Why is this name appropriate?*

d. *the annular hyperbolic plane with respect to its natural geodesic rectangular coordinate system.* (See Problem **1.8.**)

e. *the 3-manifold* $\mathbf{S}^2 \times \mathbb{R} \subset \mathbb{R}^4$*, that is the set of those points*

$$\{ (x,y,z,w) \in \mathbb{R}^4 \mid (x,y,z) \in \mathbf{S}^2 \subset \mathbb{R}^3 \}.$$

[Hint: Some of these calculations can be done with or without local coordinates. You will gain more understanding by performing the calculations more than one way.]

Appendix A

Linear Algebra from a Geometric Point of View

> Whoever thinks algebra is a trick in obtaining unknowns has thought it in vain. No attention should be paid to the fact that algebra and geometry are different in appearance. Algebras (al-jabbre and maqabeleh) are geometric facts which are proved by Propositions Five and Six of Book Two of [Euclid's] Elements.
>
> — Omar Khayyam, a paper [**A:** Khayyam (1963)]

A.0. Where Do We Start?

Usual treatments of linear and affine algebra start with a vector space as a set of "vectors" and the operations of vector addition and scalar multiplication that satisfy the axioms for a vector space. In a vector space all vectors emanate from the origin. This works well algebraically; but it ignores our geometric images and experiences of vectors.

Geometrically, we start with our experiences of Euclidean geometry where there is no point that has been singled out as the origin and where there are no numerical distances (until after a unit distance is chosen).

The linear structure of Euclidean space is carried by the translations of the space. We picture vectors as directed line segments from one point to another, and the translations serve to define when one vector is parallel to another. Any space whose translations satisfy the same properties as translations in Euclidean space is called a ***geometric affine space***, which is the subject of Section **A.1**.

The collection of all the vectors emanating from the same point is called the tangent space at that point. Using translations we may define addition and scalar multiplication of vectors. The properties of these two operations on vectors are the defining properties of a ***vector space***, which is the subject of Section **A.2**.

A.1. Geometric Affine Spaces

A ***geometric affine space over the field*** K is a space, S, together with bijections, T_{ba}: $S \to S$, $T_{ba}(a) = b$, which, for every pair of points a, b in S, satisfy the Properties (0)-(8), below. In this text K will always be the field of real numbers $\mathbb{R}$. We call T_{ba} ***the translation from*** a ***to*** b. We call the ordered pair (a,b) ***the (bound) vector from*** a ***to*** b. The most basic property of translations is:

(0) T_{ba} is unique in the sense that, if $T_{dc}(a) = b$, then $T_{ba} = T_{dc}$; and translations are closed under composition in the sense that

$$T_{ba}T_{dc} = T_{ec}, \text{ where } e = T_{ba}(d), \text{ where } (AB)(x) = A(B(x)).$$

Further, we assume that T_{aa} = identity [that is, $T_{aa}(x) = x$, for all x in S.]

[Note the implication that, if a is distinct from b, then T_{ba} has no fixed points.]

This property allows us to define when two bound vectors are equivalent. We say that (a,b) ***is parallel to*** (c,d) if there is a translation that takes (a,b) to (c,d), in symbols $(a,b)\approx(c,d)$. Property (0) assures us that this translation is unique, and, in this case, $T_{ca}(b) = d$, and thus, by (0), $T_{ca} = T_{db}$. Then we can also conclude, using (0) again, that

$$T_{ba} = T_{bd}T_{dc}T_{ca} = (T_{bd}T_{dc})T_{db} = (T_{dc}T_{bd})T_{db} = T_{dc}(T_{bd}T_{db}) = T_{dc}\,.$$

We can easily check that relation of being parallel is an equivalence relation: that is,

- $(a,b)\approx(a,b)$,
- $(a,b)\approx(c,d)$ if and only if $(c,d)\approx(a,b)$, and
- $(a,b)\approx(c,d)$ and $(a,b)\approx(e,f)$ implies that $(e,f)\approx(c,d)$.

In addition, parallel bound vectors are unique in the sense that:

- $(a,b)\approx(c,d)$ and $(a,b)\approx(c,e)$ implies that $d = e$.

This is the main property that distinguishes an affine space form other spaces. This equivalence is the same as parallel transport in Euclidean space. Parallel transport of vectors is definable in very general settings but is unique only when the space is locally isometric to Euclidean space.

We define the ***free vectors*** in S to be the equivalence classes of bound vectors. We write the equivalence class of the bound vector (a,b) to be the free vector $\mathbf{v} = [a,b]$. Note that Property (0) implies that:

- $[a,b] \leftrightarrow T_{ba}$ is a one-to-one correspondence between translations and free vectors. Thus, it follows from Property (0) that, for any point c in S,

$$(a,b) \approx (c, T_{ba}(c)).$$

Thus, every free vector $\mathbf{v}$ has a representative bound to every point c. We denote this bound vector by $\mathbf{v}_c$. We can define the addition of free vectors by

$$[c,d] + [a,b] = [c,d] + [d, T_{ba}(d)] \equiv [c, T_{ba}(d)].$$

Since $d = T_{dc}(c)$, we see that

$$[c,d] + [a,b] = [c, T_{ba}T_{dc}(c)].$$

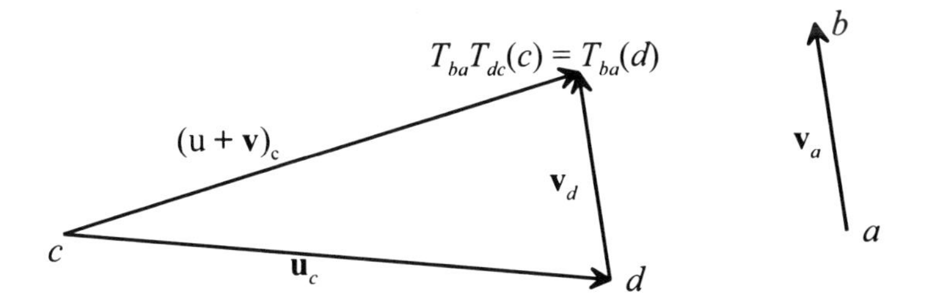

Figure A.1.1. Adding free vectors.

We can now write the further properties of a geometric affine space in terms of either free vectors or translations. For all free vectors $\mathbf{u}$, $\mathbf{v}$, $\mathbf{w}$ in S, and r, s in K:

(1) there is a free vector $\mathbf{0}$ such that $\mathbf{0} + \mathbf{v} - \mathbf{v}$;

(2) $\mathbf{u} + \mathbf{v} = \mathbf{v} + \mathbf{w}$, $[T_{ba}T_{dc} = T_{dc}T_{ba}]$;

(3) $\mathbf{u} + (\mathbf{v} + \mathbf{w}) = (\mathbf{u} + \mathbf{v}) + \mathbf{w}$, $[T_{fe}(T_{dc}T_{ba}) = (T_{fe}T_{dc})T_{ba}]$.

We now assume further that we have defined, for each r in K, a scalar multiplication of vectors, $r(a,b)$, [or an exponentiation of translations

$(T_{ba})^r$, where $r(a,b) \equiv (a,(T_{ba})^r(a))]$, with the following additional properties:

(4) $0\mathbf{v} = \mathbf{0}$, $[T_{ba}{}^0 = T_{aa}]$;

(5) $1\mathbf{v} = \mathbf{v}$, $[T_{ba}{}^1 = T_{ba}]$;

(6) $(r+s)\mathbf{v} = r\mathbf{v} + s\mathbf{v}$, $[\,(T_{ba})^{(r+s)} = (T_{ba})^r(T_{ba})^s\,]$;

{In particular, for n a positive integer,

$$n\mathbf{v} = \mathbf{v}+\mathbf{v}+...+\mathbf{v} \text{ (n times)}$$
$$[T_{ba}{}^n = T_{ba}T_{ba}...T_{ba} \text{ (n times) and } T_{ba}{}^{-1} = T_{ab}\,].\}$$

(7) $(rs)\mathbf{v} = r(s\mathbf{v})$, $[\,(T_{ba})^{(rs)} = ((T_{ba})^r)^s\,]$;

(8) $r(\mathbf{u} + \mathbf{v}) = r\mathbf{u} + r\mathbf{v}$, $[\,(T_{dc}T_{ba})^r = T_{dc}{}^rT_{ba}{}^r\,]$.

The collection of all (bound) vectors bound to a point a in S is called the ***tangent space at*** a, written

$$S_a = \{\text{ vectors } (a,b) \mid b \text{ is in } S\,\}.$$

Since each free vector $\mathbf{v} = [a,b]$ is represented by the translation T_{ba}, we can define, for each point c in S, another point:

$$\mathbf{v}(c) = [a,b](c) = T_{ba}(c)\text{; in drawings this is:}$$

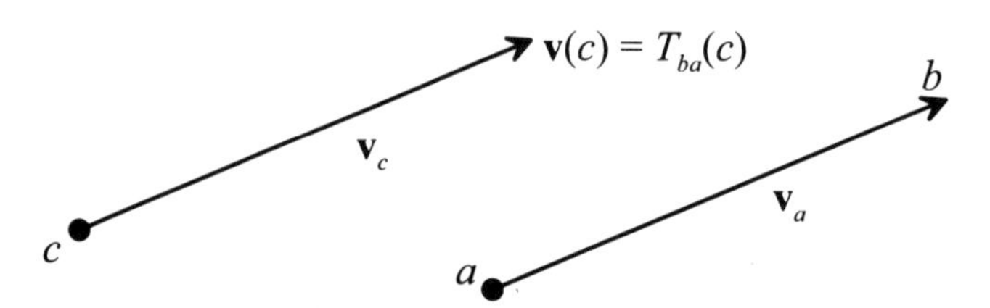

Figure A.1.2. Representing free vectors by translations.

A subset $R \subset S$ is called an ***affine subspace*** if

$$r[a,b](R) = T^r_{ba}(R) = R\ ,$$

for every pair of points a, b in R and every r in K. If $\{a_0, a_1, ..., a_n\}$ is a finite collection of points from S, then we call the ***affine span of*** $\{a_0, a_1, ..., a_n\}$, denoted by $asp\{a_0, a_1, ..., a_n\}$, the smallest affine subspace containing each of $a_0, a_1, ..., a_n$. For two points, $a \neq b$, we call $asp\{a,b\}$

the ***line determined by*** a and b. We say that $\{a_0, a_1, ..., a_n\}$ are ***affinely independent*** if, for each i,

$$a_i \text{ is not in } asp\{a_0, a_1, ...a_{i-1}, a_{i+1}, ..., a_n\}.$$

If $\{a_0, a_1, ..., a_n\}$ are affinely independent, then we say that

$$asp\{a_0, a_1, ..., a_n\}$$

is an ***n-dimensional (affine) subspace***.

Theorem A.1.1. *The vector b is in $asp\{a_0, a_1, ..., a_n\}$ if and only if*

$$b = (r_n[a_n,a_0] + ... + r_2[a_2,a_0] + r_1[a_1,a_0])(a_0)$$
$$[\text{or, } b = (T^{r_n}_{a_n a_0} \cdots T^{r_2}_{a_2 a_0} T^{r_1}_{a_1 a_0})(a_0)],$$

for some r_1 r_2 ... r_n in K.

Proof: Let R denote $asp\{a_0, a_1, ..., a_n\}$. By the definition of affine subspace,

$$T^{r_1}_{a_1 a_0}(a_0) \in R,$$

and thus,

$$(T^{r_2}_{a_2 a_0} T^{r_1}_{a_1 a_0})(a_0) = T^{r_2}_{a_2 a_0}(T^{r_1}_{a_1 a_0}(a_0)) \in R.$$

In this way we see that

$$b = (T^{r_n}_{a_n a_0} \cdots T^{r_2}_{a_2 a_0} T^{r_1}_{a_1 a_0})(a_0) = T^{r_n}_{a_n a_0}(\cdots(T^{r_2}_{a_2 a_0}(T^{r_1}_{a_1 a_0}(a_0)))\cdots) \in R.$$

Denote by R^* the collection of all points in R of the form:

$$(T^{r_n}_{a_n a_0} \cdots T^{r_2}_{a_2 a_0} T^{r_1}_{a_1 a_0})(a_0).$$

Then, clearly, R^* contains each of $a_0, a_1, ..., a_n$. We now check that R^* is an affine subspace (in which case it easily follows that $R^* = R$):

Let

$$a = (T^{s_n}_{a_n a_0} \cdots T^{s_2}_{a_2 a_0} T^{s_1}_{a_1 a_0})(a_0)$$
$$b = (T^{r_n}_{a_n a_0} \cdots T^{r_2}_{a_2 a_0} T^{r_1}_{a_1 a_0})(a_0)$$
$$c = (T^{q_n}_{a_n a_0} \cdots T^{q_2}_{a_2 a_0} T^{q_1}_{a_1 a_0})(a_0)$$

be any three points in R^*. By the properties of a geometric affine space:

$$(T^{r_n - s_n}_{a_n a_0} \cdots T^{r_2 - s_2}_{a_2 a_0} T^{r_1 - s_1}_{a_1 a_0})(a) = b,$$

and thus, by Property (0),

$$(T_{a_n a_0}^{r_n - s_n} \cdots T_{a_2 a_0}^{r_2 - s_2} T_{a_1 a_0}^{r_1 - s_1}) = T_{ba}.$$

Then

$$T_{ba}(c) = T_{ba}((T_{a_n a_0}^{q_n} \cdots T_{a_2 a_0}^{q_2} T_{a_1 a_0}^{q_1})(a_0))$$

$$= (T_{a_n a_0}^{r_n - s_n} \cdots T_{a_2 a_0}^{r_2 - s_2} T_{a_1 a_0}^{r_1 - s_1})((T_{a_n a_0}^{q_n} \cdots T_{a_2 a_0}^{q_2} T_{a_1 a_0}^{q_1})(a_0))$$

$$= (T_{a_n a_0}^{r_n - s_n + q_n} \cdots T_{a_2 a_0}^{r_2 - s_2 + q_2} T_{a_1 a_0}^{r_1 - s_1 + q_1})(a_0) \in R^*.$$

Corollary A.1.2. *If* $\{a_0, a_1, ..., a_n\}$ *are affinely independent, then the field elements* $r_1\ r_2 \ldots r_n$ *in Theorem* **A.1.1** *are unique.*

Proof: If the $r_1\ r_2 \ldots r_n$ were not unique, then

$$b = (T_{a_n a_0}^{r_n} \cdots T_{a_2 a_0}^{r_2} T_{a_1 a_0}^{r_1})(a_0) = (T_{a_n a_0}^{s_n} \cdots T_{a_2 a_0}^{s_2} T_{a_1 a_0}^{s_1})(a_0),$$

and, if i were the first index for which $r_i \neq s_i$, then

$$T_{a_i a_0}^{s_i - r_i}(a_0) = (T_{a_n a_0}^{r_n - s_n} \cdots T_{a_{i-2} a_0}^{r_{i-2}})(a_0)$$

and

$$a_i = T_{a_i a_0}(a_0) = (T_{a_i a_0}^{s_i - r_i})^{1/(s_i - r_i)}(a_0) = (T_{a_n a_0}^{r_n - s_n} \cdots T_{a_{i-2} a_0}^{r_{i-2}})^{1/(s_i - r_i)}(a_0).$$

Thus a_i would be in $asp\{a_n, ..., a_{l+1}\}$, which contradicts the hypothesis that $\{a_0, a_1, ..., a_n\}$ are affinely independent.

Theorem A.1.3. *Let* $\{a_0, a_1, ..., a_n\}$ *be any collection of affinely independent points, and* T_{ba} *be any translation, and denote* $b_i = T_{ab}(a_i)$. *Then:*

a. $\{b_0, b_1, ..., b_n\}$ *are affinely independent, and*

b. $asp\{b_0, b_1, ..., b_n\} = T_{ba}(asp\{a_0, a_1, ..., a_n\})$.

Proof. The reader can check easily that this theorem is true because

$$T_{a_i a_j}(b_j) = T_{a_i a_j}(T_{ab}(a_j)) = T_{ba}(T_{a_i a_j}(a_j)) = T_{ba}(a_i) = b_i,$$

and thus, $T_{a_i a_j} = T_{b_i b_j}$.

A.2. Vector Spaces

We can define ***vector addition*** and ***scalar multiplication*** on the tangent space S_a at the point a in S as follows:

$$(a, b) + (a, c) \equiv (a, T_{ba}(c)) = (a, T_{ba}T_{ca}(a)),$$

and

$$r(a, b) \equiv (a, T^{r}_{ab}(a)).$$

These operations satisfy the following properties, which follow from the same-numbered properties of an affine space:

(1) $\mathbf{u} + \mathbf{v} = \mathbf{v} + \mathbf{u}$;

(2) there is a vector $\mathbf{0}$ such that $\mathbf{0} + \mathbf{u} = \mathbf{u}$;

(3) $\mathbf{u} + (\mathbf{v} + \mathbf{w}) = (\mathbf{u} + \mathbf{v}) + \mathbf{w}$;

(4) $0\mathbf{u} = \mathbf{0}$;

(5) $1\mathbf{u} = \mathbf{u}$;

(6) $(r + s)\mathbf{u} = r\mathbf{u} + s\mathbf{u}$;

(7) $(rs)\mathbf{u} = r(s\mathbf{u})$; and

(8) $r(\mathbf{u} + \mathbf{v}) = r\mathbf{u} + r\mathbf{v}$.

Any set V with two binary operations satisfying these properties is called a ***vector space over the field K***.

A subset $R \subset V$ is called a ***(linear) subspace*** if $r\mathbf{u} + s\mathbf{v}$ is in R, for every pair of vectors $\mathbf{u}$, $\mathbf{v}$ in R and every r in K. If $\{\mathbf{u}_1, \mathbf{u}_2, ..., \mathbf{u}_n\}$ is a finite collection of vectors from V, then we call the ***(linear) span of*** $\{\mathbf{u}_1, \mathbf{u}_2, ..., \mathbf{u}_n\}$, denoted by $sp\{\mathbf{u}_1, \mathbf{u}_2, ..., \mathbf{u}_n\}$, the smallest (linear) subspace containing each of $\mathbf{u}_1, \mathbf{u}_2, ..., \mathbf{u}_n$. We say that $\{\mathbf{u}_1, \mathbf{u}_2, ..., \mathbf{u}_n\}$ are ***(linearly) independent*** if, for each i, $\mathbf{u}_i$ is not in

$$sp\{\mathbf{u}_1, \mathbf{u}_2, ...\mathbf{u}_{i-1}, \mathbf{u}_{i+1}, ..., \mathbf{u}_n\}.$$

If $\{\mathbf{u}_1, \mathbf{u}_2, ..., \mathbf{u}_n\}$ are linearly independent, then we say that

$$R = sp\{\mathbf{u}_1, \mathbf{u}_2, ..., \mathbf{u}_n\}$$

is an ***n-dimensional (linear) subspace*** and that $\{\mathbf{u}_1, \mathbf{u}_2, ..., \mathbf{u}_n\}$ is a ***basis*** for R.

The proofs of the two results below follow from the proofs to A.1.1. and A.1.2.

Theorem A.2.1. $\mathbf{v}$ *is in* $sp\{\mathbf{u}_1, \mathbf{u}_2, ..., \mathbf{u}_n\}$ *if and only if*

$$\mathbf{v} = r^1\mathbf{u}_1 + r^2\mathbf{u}_2 + \cdots + r^n\mathbf{u}_n,$$

for some $r^1, r^2, \ldots, r^n$ in K.

COROLLARY **A.2.2.** *If $\{\mathbf{u}_1, \mathbf{u}_2, \ldots, \mathbf{u}_n\}$ are linearly independent, then, the field elements $r^1, r^2, \ldots, r^n$ in Theorem* **A.2.1** *are unique.*

If $\{\mathbf{u}_0, \mathbf{u}_1, \ldots, \mathbf{u}_n\}$ is a ***basis*** for the subspace R, then, with respect to this basis, every **vector** v in R has a unique representation in terms of the $r_0, r_1, \ldots, r_n$. We write variously,

$$\mathbf{v} = \sum_{i=1}^{n} r^i \mathbf{u}_i = \sum r^i \mathbf{u}_i = r^i \mathbf{u}_i = (r^1, r^2, \cdots, r^n) \begin{bmatrix} \mathbf{u}_1 \\ \mathbf{u}_2 \\ \vdots \\ \mathbf{u}_n \end{bmatrix}.$$

We will usually write $\mathbf{v} = \sum r^i \mathbf{u}_i$.

We can easily check that the properties of a vector space imply that if $\mathbf{v} = \sum v^i \mathbf{u}_i$ and $\mathbf{w} = \sum w^i \mathbf{u}_i$, then

$$\mathbf{v} + \mathbf{w} = \Sigma(v^i + w^i)\mathbf{u}_i \text{ and } r\mathbf{v} = \Sigma(rv^i)\mathbf{u}_i.$$

A.3. Inner Product — Lengths and Angles

In our usual experience of Euclidean space, the notion of angle is fundamental (See Chapter 3 of [**Tx**: Henderson]), and once we have chosen a unit length, then we know how to determine the length $|\mathbf{v}|$ of any vector **v**. Assuming we know what lengths and angles are, we can define the ***Euclidean inner product*** (variously called ***the standard inner product*** or ***the dot product***) of two vectors to be:

$$\langle \mathbf{v}, \mathbf{w} \rangle = |\mathbf{v}||\mathbf{w}| \cos\theta, \text{ where } \theta \text{ is the angle between } \mathbf{v} \text{ and } \mathbf{w}.$$

We can check that this inner product satisfies the following properties:

1. ***Symmetric,*** $\langle \mathbf{v}, \mathbf{w} \rangle = \langle \mathbf{w}, \mathbf{v} \rangle$;
2. ***Bilinear,*** $r\langle \mathbf{v}, \mathbf{w} \rangle = \langle r\mathbf{v}, \mathbf{w} \rangle = \langle \mathbf{v}, r\mathbf{w} \rangle$, for all $r \in \mathrm{R}$,
   $\langle \mathbf{v} + \mathbf{u}, \mathbf{w} \rangle = \langle \mathbf{v}, \mathbf{w} \rangle + \langle \mathbf{u}, \mathbf{w} \rangle$,
   $\langle \mathbf{v}, \mathbf{u} + \mathbf{w} \rangle = \langle \mathbf{v}, \mathbf{u} \rangle + \langle \mathbf{v}, \mathbf{w} \rangle$,
3. ***Positive definite,*** $\langle \mathbf{v}, \mathbf{v} \rangle \geq 0$.

In an abstract vector space we can start by asserting the existence of ***an inner product***, which satisfies these three properties. Then we define:

$$|\mathbf{v}| = \sqrt{\langle \mathbf{v}, \mathbf{v} \rangle} \text{ and } \cos\theta = \frac{\langle \mathbf{v}, \mathbf{w} \rangle}{|\mathbf{v}||\mathbf{w}|}.$$

Note that **v** and **w** are ***perpendicular*** (or ***orthogonal***) if and only if $\langle \mathbf{v}, \mathbf{w} \rangle = 0$. The collection of all vectors orthogonal to **v** is called the ***orthogonal complement of* v**.

We can use the inner product to express projections:

The (***orthogonal***) ***projection onto the vector* a**, $P_{\mathbf{a}}(\mathbf{v})$, is defined by the following picture in $sp\{\mathbf{a},\mathbf{v}\}$.

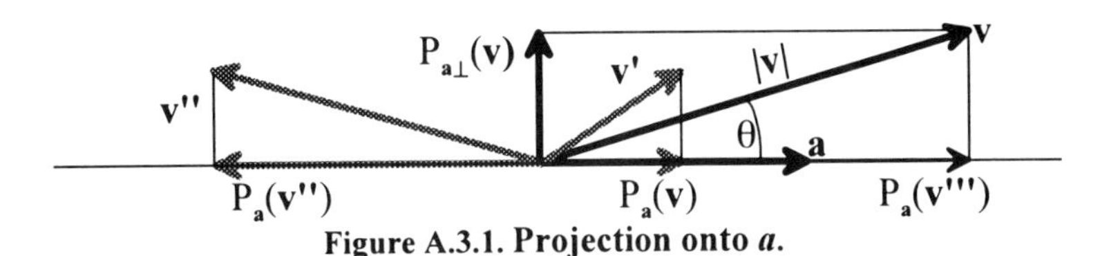

Figure A.3.1. Projection onto *a*.

Note that

$$P_{\mathbf{a}}(\mathbf{v}) = |\mathbf{v}|\cos\theta\left(\frac{1}{|\mathbf{a}|}\right)\mathbf{a},$$

where $|\mathbf{x}|$ is the length of the vector **x** and θ is the angle between **v** and **a**. Now using inner products we have:

$$P_{\mathbf{a}}(\mathbf{v}) = |\mathbf{v}|\cos\theta \frac{\mathbf{a}}{|\mathbf{a}|} = |\mathbf{v}|\frac{\langle \mathbf{v}, \mathbf{a} \rangle}{|\mathbf{v}||\mathbf{a}|}\frac{\mathbf{a}}{|\mathbf{a}|} = \frac{\langle \mathbf{v}, \mathbf{a} \rangle}{\langle \mathbf{a}, \mathbf{a} \rangle}\mathbf{a}.$$

- ***projection onto the orthogonal complement of a vector* a**:

$$P_{\mathbf{a}\perp}(\mathbf{v}) = \mathbf{v} - P_{\mathbf{a}}(\mathbf{v}) = \mathbf{v} - \frac{\langle \mathbf{v}, \mathbf{a} \rangle}{\langle \mathbf{a}, \mathbf{a} \rangle}\mathbf{a}.$$

{If the reader has not seen this before, then the reader should check carefully that this is true both formally (using the properties of the inner product) and geometrically (using Figure A.3.1).} We now apply this representation of projection to prove a famous result:

Theorem A.3.1. (***Gram-Schmidt Orthonormalization***)
If $\{\mathbf{v}_1,\mathbf{v}_2,\ldots,\mathbf{v}_n\}$ *is any basis for V, then there is another basis* $\{\mathbf{e}_1,\mathbf{e}_2,\ldots,\mathbf{e}_n\}$ *for V, such that*

a. *for each i,* $sp\{\mathbf{v}_1,\mathbf{v}_2,\ldots,\mathbf{v}_i\} = sp\{\mathbf{e}_1,\mathbf{e}_2,\ldots,\mathbf{e}_i\}$,

b. $\langle \mathbf{e}_i, \mathbf{e}_j \rangle = 0$, *for* $i \neq j$ (that is, $\mathbf{e}_i$ and $\mathbf{e}_j$ are *orthogonal*) , *and*

c. $\langle \mathbf{e}_i, \mathbf{e}_i \rangle = |\mathbf{e}_i|^2 = 1$ (that is, the $\mathbf{e}_i$ are *normalized*).

Proof: Let

$$\mathbf{e}_1 = \frac{\mathbf{v}_1}{|\mathbf{v}_1|}, \mathbf{e}_2 = \frac{\mathrm{P}_{\mathbf{e}_1\perp}(\mathbf{v}_2)}{|\mathrm{P}_{\mathbf{e}_1\perp}(\mathbf{v}_2)|} = \frac{\mathbf{v}_2 - \langle \mathbf{v}_2, \mathbf{e}_1 \rangle \mathbf{e}_1}{|\mathbf{v}_2 - \langle \mathbf{v}_2, \mathbf{e}_1 \rangle \mathbf{e}_1|}.$$

The reader should check that

$$sp\{\mathbf{v}_1\} = sp\{\mathbf{e}_1\},\ sp\{\mathbf{v}_1,\mathbf{v}_2\} = sp\{\mathbf{e}_1,\mathbf{e}_2\},$$
$$\langle \mathbf{e}_1, \mathbf{e}_2 \rangle = 0,\ \langle \mathbf{e}_1, \mathbf{e}_1 \rangle = 1, \text{ and } \langle \mathbf{e}_2, \mathbf{e}_2 \rangle = 1.$$

In general, if $\mathbf{e}_1,\mathbf{e}_2,\ldots,\mathbf{e}_k$ have been defined so that **A.3.1**.a, b, and c hold, then we can define

$$\mathbf{e}_{k+1} = \frac{\mathrm{P}_{\mathbf{e}_k\perp}(\cdots(\mathrm{P}_{\mathbf{e}_2\perp}(\mathrm{P}_{\mathbf{e}_1\perp}(\mathbf{v}_{k+1})))\cdots)}{|\mathrm{P}_{\mathbf{e}_k\perp}(\cdots(\mathrm{P}_{\mathbf{e}_2\perp}(\mathrm{P}_{\mathbf{e}_1\perp}(\mathbf{v}_{k+1})))\cdots)|} =$$
$$= \frac{\mathbf{v}_{k+1} - \langle \mathbf{v}_{k+1}, \mathbf{e}_1 \rangle \mathbf{e}_1 - \langle \mathbf{v}_{k+1}, \mathbf{e}_2 \rangle \mathbf{e}_2 - \cdots - \langle \mathbf{v}_{k+1}, \mathbf{e}_k \rangle \mathbf{e}_k}{|\mathbf{v}_{k+1} - \langle \mathbf{v}_{k+1}, \mathbf{e}_1 \rangle \mathbf{e}_1 - \langle \mathbf{v}_{k+1}, \mathbf{e}_2 \rangle \mathbf{e}_2 - \cdots - \langle \mathbf{v}_{k+1}, \mathbf{e}_k \rangle \mathbf{e}_k|}.$$

This last expression holds because the $\mathbf{e}_i$ are orthogonal and normal, and thus,

$$\mathrm{P}_{\mathbf{e}_2\perp}(\mathrm{P}_{\mathbf{e}_1\perp}(\mathbf{v})) = \mathrm{P}_{\mathbf{e}_2\perp}(\mathbf{v} - \frac{\langle \mathbf{v}, \mathbf{e}_1 \rangle}{\langle \mathbf{e}_1, \mathbf{e}_1 \rangle}\mathbf{e}_1) = \mathrm{P}_{\mathbf{e}_2\perp}(\mathbf{v} - \langle \mathbf{v}, \mathbf{e}_1 \rangle \mathbf{e}_1) =$$
$$= (\mathbf{v} - \langle \mathbf{v}, \mathbf{e}_1 \rangle \mathbf{e}_1) - \langle (\mathbf{v} - \langle \mathbf{v}, \mathbf{e}_1 \rangle \mathbf{e}_1), \mathbf{e}_2 \rangle \mathbf{e}_2 =$$
$$= (\mathbf{v} - \langle \mathbf{v}, \mathbf{e}_1 \rangle \mathbf{e}_1) - (\langle \mathbf{v}, \mathbf{e}_2 \rangle - \langle \mathbf{v}, \mathbf{e}_1 \rangle \langle \mathbf{e}_1, \mathbf{e}_2 \rangle) \mathbf{e}_2 =$$
$$= \mathbf{v} - \langle \mathbf{v}, \mathbf{e}_1 \rangle \mathbf{e}_1 - \langle \mathbf{v}, \mathbf{e}_2 \rangle \mathbf{e}_2.$$

The reader can easily check that a., b., and c. hold.

LEMMA **A.3.2**. (***The Cauchy-Schwarz Inequality***) *If* V *has an inner product, and* $\mathbf{v}, \mathbf{w} \in V$, *then* $|\langle \mathbf{v}, \mathbf{w} \rangle| \leq |\mathbf{v}||\mathbf{w}|$. *Furthermore, we have equality if and only if* $\mathbf{v}$ *and* $\mathbf{w}$ *are linearly dependent.*

Proof: If we have defined the inner product geometrically as $|\mathbf{v}||\mathbf{w}|\cos\theta$, then this lemma is trivially true from the properties of the cosine. However, if we define the inner product abstractly, then the inequality in the lemma is exactly what we need in order that the definition

$$\cos\theta = \frac{\langle \mathbf{v}, \mathbf{w} \rangle}{|\mathbf{v}||\mathbf{w}|}$$

is well defined. A proof based only on the formal properties of the inner product can be found in most linear algebra books.

A.4. Linear Transformations and Operators

A ***linear transformation*** is a function T: $V \rightarrow W$ from a vector space V to a vector space W that preserves vector addition and scalar multiplication; that is,

$$\mathrm{T}(\mathbf{v} + \mathbf{u}) = \mathrm{T}(\mathbf{v}) + \mathrm{T}(\mathbf{u}) \text{ and } \mathrm{T}(r\mathbf{v}) = r\mathrm{T}(\mathbf{v}), \text{ for all } \mathbf{v}, \mathbf{u} \in V \text{ and } r \in K.$$

The proof of the following theorem is straightforward and can be found in most linear algebra books.

Theorem A.4.1. *For any linear transformation* T: $V \rightarrow W$:

a. T(V) *is a linear subspace of* W *called* ***the image of*** T, im(T), *and the dimension of* im(T) *called the* ***rank of*** T, rank(T);

b. *the kernel*

$$\ker(\mathrm{T}) \equiv \{\mathbf{v} \in V \mid \mathrm{T}(\mathbf{v}) = \mathbf{0}\}$$

is a linear subspace of V *and its dimension is called the* ***nullity of*** T, null(T);

c. rank(T) + null(T) = dim(V);

d. *if*

(i) $\{a_1, a_2, \ldots, a_m\}$ (m = null(T)) *is a basis for* null(T),

(ii) $\{b_1, b_2, \ldots, b_r\}$ (r = rank(T)) *is a basis for* im(T), *and*

(iii) *for each* $i = 1, 2, \ldots, r$, *we pick* $c_i \in V$ *so that* $\mathrm{T}(c_i) = b_i$,

then

$$\{c_1, c_2, \ldots, c_r, a_1, a_2, \ldots, a_m\}$$

is a basis for V.

If $\{\mathbf{v}_1, \mathbf{v}_2, \ldots, \mathbf{v}_n\}$ is a basis for V and $\{\mathbf{w}_1, \mathbf{w}_2, \ldots, \mathbf{w}_m\}$ is a basis for W, then $\mathrm{T}(\mathbf{v}_i) = \Sigma\, T^j_i \mathbf{w}_j$, where the T^j_i are numbers, and then

$$\mathrm{T}(\mathbf{a}) = \mathrm{T}(\Sigma\, a^i\mathbf{v}_i) = \Sigma\, a^i\mathrm{T}(\mathbf{v}_i) = \Sigma\, a^i(\Sigma\, T_i^j\mathbf{w}_j) = \Sigma\,\Sigma(a^iT_i^j)\mathbf{w}_j.$$

The numbers T_i^j form an n by m matrix, called the ***matrix of*** T ***with respect to the bases*** $\{\mathbf{v}_1,\mathbf{v}_2,\ldots,\mathbf{v}_n\}$ ***and*** $\{\mathbf{w}_1,\mathbf{w}_2,\ldots,\mathbf{w}_m\}$. For different bases there would be different matrices. And, conversely, with respect to these two bases any n by m matrix (M_i^j) will determine a linear transformation, M, by setting

$$M(\Sigma\, a^i\mathbf{v}_i) = \Sigma\,\Sigma(a^iM_i^j)\mathbf{w}_j.$$

The reader can easily check that, in Theorem **A.4.1**, with respect to the basis

$$\{c_1,c_2,\ldots,c_r,a_1,a_2,\ldots,a_n\} \text{ for } V$$

and any basis

$$\{b_1,b_2,\ldots,b_r,d_1,d_2,\ldots,d_s\} \text{ for } W,$$

we have the following corollary:

Corollary A.4.2. *For any linear transformation* T: $V\to W$, *there are bases for V and W such that with respect to these bases,* T *is represented by a matrix with* r (= rank(T)) 1*'s on the diagonal and all the other entries zero.*

If $V = W$, then a linear transformation T: $V\to V$ is called a ***linear operator***.

Examples of linear operators from $\mathbb{R}^n$ to $\mathbb{R}^n$ are:

- ***Dilation by*** λ: $\mathrm{M}_\lambda(\mathbf{v}) = \lambda\mathbf{v}$.
- ***Projection onto a vector*** **a**: (See Section **A.3**.)

$$\mathrm{P}_{\mathbf{a}}(\mathbf{v}) = |\mathbf{v}|\cos\theta\,\frac{\mathbf{a}}{|\mathbf{a}|} = \frac{\langle\mathbf{v},\mathbf{a}\rangle}{\langle\mathbf{a},\mathbf{a}\rangle}\,\mathbf{a}$$

The fact that

$$\mathrm{P}_{\mathbf{a}}(\mathbf{v}+\mathbf{w}) = \mathrm{P}_{\mathbf{a}}(\mathbf{v}) + \mathrm{P}_{\mathbf{a}}(\mathbf{w})$$

can be seen geometrically by making a model (in your mind) of the subspace $sp\{\mathbf{a},\mathbf{v},\mathbf{w}\}$.

- ***Reflection through the orthogonal complement of a vector*** **a**:

$$F_{\mathbf{a}\perp}(\mathbf{v}) = \mathbf{v} - 2P_{\mathbf{a}}(\mathbf{v}) = P_{\mathbf{a}\perp}(\mathbf{v}) - P_{\mathbf{a}}(\mathbf{v}).$$

- ***Projection onto the orthogonal complement of a vector* a**:

$$P_{\mathbf{a}\perp}(\mathbf{v}) = \mathbf{v} - P_{\mathbf{a}}(\mathbf{v}).$$

- ***Dilation by* λ *in the direction of* a**:

$$M_{\lambda\mathbf{a}}(\mathbf{v}) = P_{\mathbf{a}\perp}(\mathbf{v}) + \lambda P_{\mathbf{a}}(\mathbf{v}).$$

  Note that if $\lambda = 0$, then this dilation is just a projection onto the orthogonal complement of **a**; and if $\lambda = -1$, then it is a reflection through the orthogonal complement of **a**.

- For two orthogonal vectors, $\langle \mathbf{w}, \mathbf{a} \rangle = 0$, a **(w,a)**-***shear*** is the linear operator

$$S_{\mathbf{w},\mathbf{a}}(\mathbf{v}) = \mathbf{v} + \langle \mathbf{v}, \mathbf{w} \rangle \mathbf{a} \ .$$

  The reader should check that a **(w,a)**-shear takes **a** to **a** and preserves planes that are orthogonal to **w**. [Hint: Check that $\langle S_{\mathbf{w},\mathbf{a}}(\mathbf{v}), \mathbf{w} \rangle = \langle \mathbf{v}, \mathbf{w} \rangle$.]

- ***Rotation in the plane of* a *and* b *through an angle* θ**:

$$R_{\theta,\mathbf{ab}}(\mathbf{v}) = \mathbf{v} - [P_{\mathbf{a}}(\mathbf{v}) + P_{\mathbf{b}}(\mathbf{v})] + R_{\theta}[P_{\mathbf{a}}(\mathbf{v}) + P_{\mathbf{b}}(\mathbf{v})],$$

  where R_θ is the ordinary rotation through angle θ about **0** in plane $as\{\mathbf{a},\mathbf{b}\}$.

- ***The sum or product of linear operators is a linear operator***: that is, if T and S are linear operators, then T + S and TS are also linear operators, where

$$(T + S)(\mathbf{v}) = T(\mathbf{v}) + S(\mathbf{v}) \text{ and } (TS)(\mathbf{v}) = T(S(\mathbf{v})).$$

THEOREM A.4.3. *Every linear operator from $\mathbb{R}^n$ to $\mathbb{R}^n$ over the field of reals $\mathbb{R}$ is the composition of a finite number of shears, reflections, and dilations.*

Outline of a geometric proof—an algebraic proof will be given at the end of this section: Let T be any linear operator and use Theorem

A.4.1 followed by the Gramm-Schmidt Orthonormalization (**A.3.1**) to find an orthonormal basis $\{\mathbf{e}_1,\mathbf{e}_2,...,\mathbf{e}_n\}$ for $\mathbb{R}^n$, such that

$$\mathscr{B} = \{T(\mathbf{e}_1),T(\mathbf{e}_2),...,T(\mathbf{e}_r)\}$$

is a basis for $T(\mathbb{R}^n)$ and $\{\mathbf{e}_{r+1},\mathbf{e}_{r+2},...,\mathbf{e}_n\}$ is a basis for null(T), where $r = n\text{-}m$ is the rank of T (that is, the dimension of $T(\mathbb{R}^n)$). Since T is determined by the n vectors $\mathscr{B}$, we will have proved the theorem if we show that there is a composition of a finite number of projections, shears, reflections, dilations, and rotations that takes $\{\mathbf{e}_1,\mathbf{e}_2,...,\mathbf{e}_n\}$ to $\mathscr{B}$. Since $\mathscr{B}$ is a basis for $T(\mathbb{R}^n)$, we can use the proof of **A.3.1** to find an orthonormal basis $\{\mathbf{b}_1,\mathbf{b}_2,...,\mathbf{b}_r\}$ for $T(\mathbb{R}^n)$ satisfying the conclusions (a), (b), and (c) of the theorem. We will use this basis $\{\mathbf{b}_1,\mathbf{b}_2,...,\mathbf{b}_r\}$ in our proof.

First, if $\mathbf{e}_1 \neq \mathbf{b}_1$ then we can reflect (through the orthogonal complement of $\mathbf{e}_1 - \mathbf{b}_1$) to take

$$\mathbf{e}_1 \text{ to } \mathbf{b}_1 = T(\mathbf{e}_1)/|T(\mathbf{e}_1)|$$

and then dilate by $|T(\mathbf{e}_1)|$ in the direction of $T(\mathbf{e}_1)$. The result will be an operator A_1 that is the composition of a reflection and a dilation, such that $A_1(\mathbf{e}_1) = T(\mathbf{e}_1)$ and such that

$$\{A_1(\mathbf{e}_1),A_1(\mathbf{e}_2),...,A_1(\mathbf{e}_n)\}$$

is an orthogonal basis with $|A_1(\mathbf{e}_i)| = 1$, for $i > 1$.

Second, if $A_1(\mathbf{e}_2) \neq \mathbf{b}_2$, then reflect through the orthogonal complement of $A_1(\mathbf{e}_2) - \mathbf{b}_2$ to take

$$A(\mathbf{e}_2) \text{ to } \mathbf{b}_2 = \frac{P_{\mathbf{b}_1\perp}(T(\mathbf{e}_2))}{|P_{\mathbf{b}_1\perp}(T(\mathbf{e}_2))|},$$

then dilate by $|P_{\mathbf{b}_1\perp}(T(\mathbf{e}_2))|$ in the direction of $P_{\mathbf{b}_1\perp}(T(\mathbf{e}_2))$, and finally perform a $(\lambda\mathbf{b}_2,\mathbf{b}_1)$-shear to take $P_{\mathbf{b}_1\perp}(T(\mathbf{e}_2))$ to $T(\mathbf{e}_1)$. (See Figure A.4.2.)

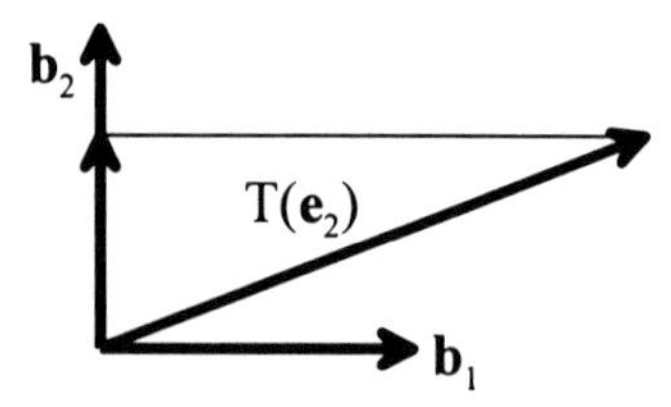

Figure A.4.2. Second step in the proof of A.4.3.

The reader can check that

$$\lambda = \frac{|P_{\mathbf{b}_1}(T(\mathbf{e}_2))|}{|P_{\mathbf{b}_1\perp}(T(\mathbf{e}_2))|}$$

and thus is positive. Note that this rotation is a rigid motion, and that the dilation and shear do not change any of the images of $\{\mathbf{e}_1,\mathbf{e}_3,\ldots,\mathbf{e}_n\}$.

Thus, after the second stage, we have a composition, A_2, of rotations, shears, and dilations such that

$$A_2(\mathbf{e}_1) = T(\mathbf{e}_1),\ A_2(\mathbf{e}_2) = T(\mathbf{e}_2),\ \{A_2(\mathbf{e}_1),A_2(\mathbf{e}_2),\ldots,A_2(\mathbf{e}_n)\}$$

is an orthogonal basis, and $A_2(\mathbf{e}_i) = 1$, for $i > 2$. The interested reader can now see how to continue.

The reader can check that with respect to the orthonormal basis $\{\mathbf{e}_1,\mathbf{e}_2,\ldots,\mathbf{e}_n\}$, we have the following matrices:

- ***Scaling (magnifying) by λ:***

$$\text{matrix}(M_\lambda) = \begin{pmatrix} \lambda & 0 & \cdots & 0 & 0 \\ 0 & \lambda & \cdots & 0 & 0 \\ \vdots & \vdots & \ddots & \vdots & \vdots \\ 0 & 0 & \cdots & \lambda & 0 \\ 0 & 0 & \cdots & 0 & \lambda \end{pmatrix} = \lambda I.$$

- ***Projection onto the vector*** $\mathbf{e}_k$:

$$\text{matrix}(P_{\mathbf{e}_k}) = \text{0-matrix except } k \rightarrow \begin{pmatrix} \ddots & \vdots & \vdots & \vdots & \ddots \\ \cdots & 0 & 0 & 0 & \cdots \\ \cdots & 0 & 1 & 0 & \cdots \\ \cdots & 0 & 0 & 0 & \cdots \\ \ddots & \vdots & \vdots & \vdots & \ddots \end{pmatrix}.$$

- ***Reflection through the orthogonal complement of*** $\mathbf{e}_k$:

$$\text{matrix}(F_{\mathbf{e}_k\perp}) = \text{I-matrix except } k \rightarrow \begin{pmatrix} \ddots & \vdots & \cdots & \vdots & \ddots \\ \cdots & 1 & 0 & 0 & \cdots \\ \cdots & 0 & -1 & 0 & \cdots \\ \cdots & 0 & 0 & 1 & \cdots \\ \ddots & \vdots & \vdots & \vdots & \ddots \end{pmatrix}.$$

- ***Reflection through the orthogonal complement of*** $\mathbf{e}_k$-$\mathbf{e}_l$**:**

$$\text{matrix}(F_{(\mathbf{e}_k-\mathbf{e}_l)\perp}) = \text{I-matrix except} \begin{array}{r} \\ k \rightarrow \\ \\ l \rightarrow \\ \\ \end{array} \begin{pmatrix} \ddots & \vdots & \cdots & \vdots & \ddots \\ \cdots & 0 & \cdots & 1 & \cdots \\ \vdots & \vdots & \ddots & \vdots & \vdots \\ \cdots & 1 & \cdots & 0 & \cdots \\ \ddots & \vdots & \cdots & \vdots & \ddots \end{pmatrix}.$$

Note that multiplying by this matrix (on the left) is the same as the *elementary row operation* of interchanging the k-th and l-th rows.

- ***Projection onto the orthogonal complement of*** $\mathbf{e}_k$:

$$\text{matrix}(P_{\mathbf{e}_k\perp}) = \text{I-matrix except} \begin{array}{r} \\ \\ k \rightarrow \\ \\ \\ \end{array} \begin{pmatrix} \ddots & \vdots & \vdots & \vdots & \ddots \\ \cdots & 1 & 0 & 0 & \cdots \\ \cdots & 0 & 0 & 0 & \cdots \\ \cdots & 0 & 0 & 1 & \cdots \\ \ddots & \vdots & \vdots & \vdots & \ddots \end{pmatrix}.$$

- ***Dilation by*** λ ***in the direction of*** $\mathbf{e}_k$:

$$\text{matrix}(M_{\lambda\mathbf{e}_k}) = \text{I-matrix except} \begin{array}{r} \\ \\ k \rightarrow \\ \\ \\ \end{array} \begin{pmatrix} \ddots & \vdots & \vdots & \vdots & \ddots \\ \cdots & 1 & 0 & 0 & \cdots \\ \cdots & 0 & \lambda & 0 & \cdots \\ \cdots & 0 & 0 & 1 & \cdots \\ \ddots & \vdots & \vdots & \vdots & \ddots \end{pmatrix}.$$

Note that multiplying by this matrix (on the left) is the same as multiplying the k-th row by λ.

- $(\mathbf{e}_k, \lambda\mathbf{e}_l)$-***shear***:

$$\text{matrix}(S_{\mathbf{e}_k,\lambda\mathbf{e}_l}) = \text{I-matrix except} \begin{array}{r} \\ k \rightarrow \\ \\ l \rightarrow \\ \\ \end{array} \begin{pmatrix} \ddots & \vdots & \cdots & \vdots & \ddots \\ \cdots & 1 & \cdots & 0 & \cdots \\ \vdots & \vdots & \ddots & \vdots & \vdots \\ \cdots & \lambda & \cdots & 1 & \cdots \\ \ddots & \vdots & \cdots & \vdots & \ddots \end{pmatrix}.$$

Note that multiplying by this matrix (on the left) is the same as the *elementary row operation* of adding to the k-th row λ times the l-th row.

- ***Rotation in the plane of*** $\mathbf{e}_k$ ***and*** $\mathbf{e}_l$ ***through an angle*** θ:

$$\text{matrix}(R_{\theta,\mathbf{e}_k\mathbf{e}_l}) = \text{I-matrix except} \quad \begin{matrix} \\ k \to \\ \\ l \to \\ \\ \end{matrix} \begin{pmatrix} \ddots & \vdots & \cdots & \vdots & \ddots \\ \cdots & \cos\theta & \cdots & \sin\theta & \cdots \\ \vdots & \vdots & \ddots & \vdots & \vdots \\ \cdots & -\sin\theta & \cdots & \cos\theta & \cdots \\ \ddots & \vdots & \cdots & \vdots & \ddots \end{pmatrix}.$$

- ***The sum or product of linear transformations is linear***: that is, if T and S are linear transformations with matrices T^j_i and S^j_i, then

$$(T + S)^j_i = (T^j_i + S^j_i) \text{ and } (T(S))^j_i = \Sigma T^j_k S^k_i .$$

Alternate proof of Theorem **A.4.3**: If (T) is the matrix of the linear operator with respect to some orthonormal basis, the matrix (T) can be reduced to a diagonal matrix by a finite number of the types of row (or column) operations:

- interchanging the k-th and l-th rows (columns), which is the same as multiplying on the left (right) by $\text{matrix}(F_{(\mathbf{e}_k-\mathbf{e}_l)\perp})$,
- adding to the k-th row (column) λ times the l-th row (column), which is the same as multiplying on the left (right) by $\text{matrix}(S_{\mathbf{e}_k,\lambda\mathbf{e}_l})$.

Thus, we have $E(T) = D$ (or, $(T)E = D$), where E is a finite product of reflection or shear matrices and D is a diagonal matrix. But each of the reflection and shear matrices has an inverse, which is of the same type, and any diagonal matrix is the product of matrices that

- multiply one row by a scalar λ, which is the same as multiplying by $\text{matrix}(M_{\lambda\mathbf{e}_k})$.

Thus, we can write

$$(T) = E^{-1}M \text{ (or, } ME^{-1}),$$

where now the right hand side is a finite product of reflections, shears, and dilations.

A.5. Areas, Cross Products, and Triple Products

DEFINITION: The ***cross product*** $\mathbf{v} \times \mathbf{w}$ of two vectors $\mathbf{v}$ and $\mathbf{w}$ in $\mathbb{R}^3$ is the vector

1. whose magnitude is the area of the parallelogram formed by $\mathbf{v}$ and $\mathbf{w}$,
2. which is perpendicular to both $\mathbf{v}$ and $\mathbf{w}$, and
3. whose direction is such that $\mathbf{v}$, $\mathbf{w}$, and $\mathbf{v} \times \mathbf{w}$ (in this order) form a right-hand system. (If you curl the fingers of your right hand from $\mathbf{v}$ to $\mathbf{w}$, then your thumb points in the direction of $\mathbf{v} \times \mathbf{w}$.)

Note that if the magnitude of $\mathbf{v} \times \mathbf{w}$ is 0, then the direction of $\mathbf{v} \times \mathbf{w}$ is not defined, which is correct in this case, because then $\mathbf{v} \times \mathbf{w} = \mathbf{0}$.

THEOREM A.5.1. *For any two vectors* $\mathbf{v}$ *and* $\mathbf{w}$ *in* $\mathbb{R}^3$*, we have:*

a. $\mathbf{v} \parallel \mathbf{w}$ *if and only if* $\mathbf{v} \times \mathbf{w} = \mathbf{0}$,

b. $|\mathbf{v} \times \mathbf{w}| = |\mathbf{v}|\,|\mathbf{w}| \sin \theta_{\mathbf{v},\mathbf{w}} = \sqrt{|\mathbf{v}|^2|\mathbf{w}|^2 - \langle \mathbf{v}, \mathbf{w} \rangle^2}$,

c. $\mathbf{v} \times \mathbf{w}$ *is bilinear* $[\lambda(\mathbf{v} \times \mathbf{w}) = (\lambda\mathbf{v}) \times \mathbf{v} = \mathbf{v} \times (\lambda\mathbf{w})$
$(\mathbf{u} + \mathbf{v}) \times \mathbf{w} = (\mathbf{u} \times \mathbf{w}) + (\mathbf{v} \times \mathbf{w})$
$\mathbf{v} \times (\mathbf{w} + \mathbf{u}) = (\mathbf{v} \times \mathbf{w}) + (\mathbf{v} \times \mathbf{u})]$, *and*

d. $\mathbf{v} \times \mathbf{w}$ *is anticommutative* $[\mathbf{v} \times \mathbf{w} = -(\mathbf{w} \times \mathbf{v})]$.

Proof: Parts **a**, **b**, and **d** follow immediately from the definition and (for **b**) the geometric definition of $\langle \mathbf{v}, \mathbf{w} \rangle$. To see part **c**, check that the function

$$\mathbf{x} \rightarrow \mathbf{v} \times \mathbf{x} \text{ is equal to } R \circ P_{\mathbf{v}\perp} ,$$

where $P_{\mathbf{v}\perp}$ is the projection onto the plane orthogonal to $\mathbf{v}$, and R is the $\pi/2$-rotation in the plane orthogonal to $\mathbf{v}$ in the direction counterclockwise from the point-of-view of $\mathbf{v}$. The linearity of the cross product now follows because both R and $P_{\mathbf{v}\perp}$ are linear.

DEFINITION: The ***triple product*** (or ***box product***) $[\mathbf{uvw}]$ of three vectors in $\mathbb{R}^3$ is defined as a number whose

- absolute value is the volume of the parallelepiped determined by the three vectors, and

- sign is positive (negative) if **u**, **v**, **w** forms a right (left) hand system.

Theorem A.5.2. *For any three vectors* **u**, **v**, *and* **w** *in* $\mathbb{R}^3$, *we have*:

a. $[\mathbf{uvw}] = [\mathbf{vwu}] = [\mathbf{wuv}] = -[\mathbf{uvw}] = -[\mathbf{vwu}] = -[\mathbf{wuv}]$.

b. $[\mathbf{uvw}] = \langle \mathbf{u} \times \mathbf{v}, \mathbf{w}\rangle = \langle \mathbf{u}, \mathbf{v} \times \mathbf{w}\rangle = \langle \mathbf{w} \times \mathbf{u}, \mathbf{v}\rangle$.

c. $[\mathbf{uvw}] = 0$ *if and only if the three vectors are linearly independent.*

d. *If* $\{\mathbf{e}_1,\mathbf{e}_2,\mathbf{e}_3\}$ *is a right-handed orthonormal basis for* $\mathbb{R}^3$, *then with respect to this basis,*

$$[\mathbf{u},\mathbf{v},\mathbf{w}] = \langle \mathbf{u}\times\mathbf{v},\mathbf{w}\rangle = \det\begin{pmatrix} \mathbf{u}^1 & \mathbf{v}^1 & \mathbf{w}^1 \\ \mathbf{u}^2 & \mathbf{v}^2 & \mathbf{w}^2 \\ \mathbf{u}^3 & \mathbf{v}^3 & \mathbf{w}^3 \end{pmatrix},$$

where

$$\mathbf{u} = \Sigma \mathbf{u}^i \mathbf{e}_i,\ \mathbf{v} = \Sigma \mathbf{v}^i \mathbf{e}_i,\ \textit{and}\ \mathbf{w} = \Sigma \mathbf{w}^i \mathbf{e}_i.$$

Proof: The reader should be able to check parts **a**, **b**, **c** directly from the definition of $[\mathbf{uvw}]$ and the geometric definition of the inner product. Part **d** follows directly from the definition of determinant in **A.6** or can at this point be taken as a definition of the determinant in $\mathbb{R}^3$. Or more directly, it is easy to check that elementary column operations will not change the volume.

Theorem A.5.3. (Double Cross Formula) *For any three vectors* **u**, **v**, *and* **w** *in* $\mathbb{R}^3$ *we have*:

$$(\mathbf{u} \times \mathbf{v}) \times \mathbf{w} = \langle \mathbf{u}, \mathbf{w}\rangle\mathbf{v} - \langle \mathbf{v}, \mathbf{w}\rangle\mathbf{u}$$

and thus

$$\mathbf{w} \times (\mathbf{u} \times \mathbf{v}) = \langle \mathbf{v},\mathbf{w}\rangle\mathbf{u} - \langle \mathbf{u}, \mathbf{w}\rangle\mathbf{v}.$$

Proof: In the special case that **u** equals **w**, $(\mathbf{u} \times \mathbf{v}) \times \mathbf{u}$ is in the same direction as $P_{\mathbf{u}\perp}(\mathbf{v})$–to see this, use your right hand–and its magnitude is $|\mathbf{u}|^2|P_{\mathbf{u}\perp}(\mathbf{v})|$. Thus,

$$(\mathbf{u}\times\mathbf{v})\times\mathbf{u} = |\mathbf{u}|^2 P_{\mathbf{u}\perp}(\mathbf{v}) = |\mathbf{u}|^2\left[\mathbf{v} - \frac{\langle \mathbf{u},\mathbf{v}\rangle}{|\mathbf{u}|^2}\mathbf{u}\right] = \langle \mathbf{u},\mathbf{u}\rangle\mathbf{v} - \langle \mathbf{u},\mathbf{v}\rangle\mathbf{u}.$$

If $\mathbf{u}$ and $\mathbf{v}$ are parallel, then both sides are zero. If $\mathbf{u}$ and $\mathbf{v}$ are not parallel, then $\{\mathbf{u}, \mathbf{v}, \mathbf{u} \times \mathbf{v}\}$ is a basis for $\mathbb{R}^3$, and thus, $\mathbf{w}$ is a linear combination:

$$\mathbf{w} = r\mathbf{u} + s\mathbf{v} + t(\mathbf{u} \times \mathbf{v}).$$

Therefore,

$$\begin{aligned}
(\mathbf{u} \times \mathbf{v}) \times \mathbf{w} &= (\mathbf{u} \times \mathbf{v}) \times (r\mathbf{u} + s\mathbf{v} + t(\mathbf{u} \times \mathbf{v})) = \\
&= r((\mathbf{u} \times \mathbf{v}) \times \mathbf{u}) + s((\mathbf{u} \times \mathbf{v}) \times \mathbf{v}) + t((\mathbf{u} \times \mathbf{v}) \times (\mathbf{u} \times \mathbf{v})) = \\
&= r((\mathbf{u} \times \mathbf{v}) \times \mathbf{u}) - s((\mathbf{v} \times \mathbf{u}) \times \mathbf{v}) + 0 = \\
&= r\langle \mathbf{u}, \mathbf{u}\rangle \mathbf{v} - r\langle \mathbf{v}, \mathbf{u}\rangle \mathbf{u} - s\langle \mathbf{v}, \mathbf{v}\rangle \mathbf{u} + \langle \mathbf{v}, \mathbf{u}\rangle \mathbf{v} = \\
&= \langle r\mathbf{u}+s\mathbf{v}, \mathbf{u}\rangle \mathbf{v} - \langle r\mathbf{u}+s\mathbf{v}, \mathbf{v}\rangle \mathbf{v} = \\
&= \langle r\mathbf{u}+s\mathbf{v}+t(\mathbf{u} \times \mathbf{v}), \mathbf{u}\rangle \mathbf{v} - \langle r\mathbf{u}+s\mathbf{v}+t(\mathbf{u} \times \mathbf{v}), \mathbf{v}\rangle \mathbf{v} = \\
&= \langle \mathbf{u}, \mathbf{w}\rangle \mathbf{v} - \langle \mathbf{v}, \mathbf{w}\rangle \mathbf{u}.
\end{aligned}$$

A.6. Volumes, Orientation, and Determinants

We now examine the effect of a linear operator on volumes in higher dimensions. The usual approach in linear algebra books is to start with the algebra and then go to the geometry. They define the determinant of a square matrix, and then show that all the matrices that represent the same linear operator have the same determinant, which can then be called the determinant of the linear operator. It is then shown that if T is a linear operator from $\mathbb{R}^n$ to $\mathbb{R}^n$, then this determinant is equal to

$$\frac{n\text{-volume of T}(C)}{n\text{-volume of } C}$$

for any n-cube C in $\mathbb{R}^n$. We will proceed in the reverse direction. We will start with the geometry and proceed to the algebra.

THEOREM A.6.1. *For every linear operator* T *from* $\mathbb{R}^n$ *to* $\mathbb{R}^n$ *and every n-cube C in* $\mathbb{R}^n$, *the ratio*

$$\frac{|n\text{-volume of T}(C)|}{|n\text{-volume of } C|}$$

is a constant independent of C.

Proof: Since T is the product of reflections, shears, and dilations (Theorem **A.4.3**), we need only prove the theorem for these 3 types of operators. The reader can check that reflections and shears do not change volumes and that a dilation of λ in the direction of a vector **V** changes volumes by the ratio of λ. This completes the proof.

DEFINITION: We denote the constant from **A.6.1** by $|\det(T)|$. If you are in a setting (such as $\mathbb{R}^2$ or $\mathbb{R}^3$) where there is a clear understanding of orientation then you can give $\det(T)$ a sign by declaring that

$$\det(T) = |\det(T)|, \text{ if } C \text{ and } T(C) \text{ have the same orientation,}$$

and

$$\det(T) = -|\det(T)|, \text{ if } C \text{ and } T(C) \text{ have different orientations.}$$

We will return to the issue of orientation later (**A.6.4** and **A.6.5**).

THEOREM A.6.2. *If* T *and* S *are two linear operators from* $\mathbb{R}^n$ *to* $\mathbb{R}^n$, *then*

$$|\det(TS)| = |\det(T)|\,|\det(S)|.$$

Proof. We see that

$$|\det(TS)| = \frac{|n\text{-volume of } TS(C)|}{|n\text{-volume of } C|} =$$

$$= \frac{|n\text{-volume of } T(S(C))|}{|n\text{-volume of } S(C)|}\,\frac{|n\text{-volume of } S(C)|}{|n\text{-volume of } C|} =$$

$$= \frac{|n\text{-volume of } T(S(C))|}{|n\text{-volume of } S(C)|}\,|\det(S)|.$$

So the proof will be completed if we show that

$$\det(T) = \frac{|n\text{-volume of } T(S(C))|}{|n\text{-volume of } S(C)|}.$$

But n-volume can be calculated by filling the region with smaller and smaller cubes and then taking a limit. Thus, the ratio of the areas of the little cubes and their images will be $|\det(T)|$ and so also the limit.

COROLLARY A.6.3. *The following are equivalent:*

a. $|\det(T)| \neq 0$.

b. T *takes any basis to another basis.*

c. T *has an inverse* T^{-1} *such that* $TT^{-1} = T^{-1}T = \text{identity}$, *and*

$$|\det(T^{-1})| = \frac{1}{|\det(T)|}.$$

DEFINITION: If M is a matrix that represents the linear operator T with respect to a basis $\mathscr{B}$, then we define $|\det(M)| = |\det(T)|$. This definition is well defined because the *Alternate proof of Theorem* **A.4.3** shows that $|\det(S)|$ and $|\det(T)|$ are both determined in the same way by their common matrix M, using either row or column operations.

THEOREM A.6.4. *If* T *is a linear operator* T *with* $|\det(T)| \neq 0$, *then there is a one-parameter family of linear operators* T_t $(0 \le t \le 1)$ *such that*

a. $T_0 = T$,

b. *for all* t, $|\det(T_t)| \neq 0$,

c. *the function* $t \to M_t$ *is continuous where* M_t *is the matrix for* T_t *with respect to a fixed basis.*

d. T_1 *is either the identity operator or a reflection through the orthogonal complement of a fixed vector* **V**.

Any one-parameter family satisfying **a-c** is called an ***isotopy of*** T.

Proof: First note that the theorem is true if T is either a shear, dilation, or reflection, and in the case of shears and dilations (with $\lambda = 1$), T_1 is the identity. If the theorem is true for operators T and S, then T_tS_t is easily seen to be an isotopy of TS with $(TS)_1 = T_1S_1$ equal either to the identity or to a reflection or to the product of two reflections. The product of two reflections is a rotation, which is clearly isotopic to the identity. The theorem now follows because every linear operator is the product of dilations, shears, and reflections.

DEFINITION: If T_1 is the identity we say that T is ***orientation preserving***. If T_1 is a reflection we say that T is ***orientation reversing***.

THEOREM A.6.5. *If* T *is a linear operator with* $|\det(T)| \neq 0$, *then* T *cannot be both orientation preserving and orientation reversing. Thus, we can define*

$$\det(T) = |\det(T)|, \textit{ if } T \textit{ is orientation preserving}$$

$$\det(T) = -|\det(T)|, \textit{ if } T \textit{ is orientation reversing.}$$

Proof: In dimensions 2 and 3 this is geometrically obvious because there is a clear meaning of orientation, and we know that a right-hand system cannot be isotoped to a left-hand system. Dimensions 2 and 3 are the only dimensions that we need in this book. In higher dimensions, the geometry is not so obvious, and thus, one may wish to resort to the algebraic proof that there is a unique function defined on n-tuples of n-vectors, which is

1. ***multilinear***—linear in each (vector) variable,
2. ***alternating***—if you interchange two vectors, you change the sign (note that this corresponds to a reflection), and
3. ***normed***—equal to 1 on the identity matrix.

See [**LA**: Damiano] for a proof of this.

A.7. Eigenvalues and Eigenvectors

A nonzero vector **v** is called an ***eigenvector*** for the linear operator T if

$$T(\mathbf{v}) = \lambda\mathbf{v}, \text{ for some scalar } \lambda.$$

The scalar is called the ***eigenvalue*** associated with the eigenvector **v**. Note that if **v** is an eigenvector, then so is $a\mathbf{v}$, for and $a \neq 0$.

THEOREM **A.7.1**. *The following statements about the linear operator* T *are equivalent:*

a. T *has an inverse,*

b. $\ker(T) = 0$,

c. $\det(T) \neq 0$,

d. 0 *is **not** an eigenvalue for* T.

Proof: T has an inverse if and only if T is one-to-one and onto, which is true (by **A.4.1**) if and only if $\ker(T) = 0$. T has an inverse if and only if (by **A.6.3**) $\det(T) \neq 0$. But, if T has 0 as an eigenvalue, then the associated eigenvector is in the kernel ker(T), and if **v** is a nonzero vector in ker(T), then $T(\mathbf{v}) = \mathbf{0} = 0\mathbf{v}$, and thus, 0 is an eigenvalue.

Thus, if λ is an eigenvalue for T, and **v** is its associated eigenvector, then $T(\mathbf{v})-\lambda\mathbf{v} = \mathbf{0}$ and **v** is in the kernel of $T-\lambda I$, and thus, we have:

THEOREM A.7.2. *For a linear operator* T, λ *is an eigenvalue if and only if*

$$\det(T-\lambda I) = 0.$$

Once we have found the eigenvalues, then the eigenvectors can be found by solving the linear equation $T(\mathbf{v})-\lambda\mathbf{v} = \mathbf{0}$ for **v**.

THEOREM A.7.3. (The Principal Axis Theorem). *If* T *is a* ***symmetric*** *linear operator on* $\mathbb{R}^n$, *that is, for every* **v**, **w**,

$$\langle T(\mathbf{v}),\mathbf{w}\rangle = \langle \mathbf{v},T(\mathbf{w})\rangle,$$

then there is an orthonormal basis, $\{\mathbf{e}_1,\ldots,\mathbf{e}_n\}$, *for* $\mathbb{R}^n$, *consisting entirely of eigenvectors of* T *with real eigenvalues.*

Proofs can be found in many linear algebra texts.

A.8. Introduction to Tensors

We do not use the language of tensors significantly in this book, but many of the notions in this book can be described in the tensor language. In this section we will introduce the terminology of tensors that is used in many treatments of differential geometry, and then we will use this terminology to describe many of the notions used in this book.

If V is a vector space over the field K, then a linear function from V to K is called a ***linear functional***.

Examples of linear functionals:

1. The i-th coordinate functional: $\mathbf{v} \to v^i$, which assigns to each vector **v** its i-th coordinate with respect some fixed basis.

2. The directional derivative of f with respect to a tangent vector: $\mathbf{X}_p \to \mathbf{X}_p f$, which assigns to each vector $\mathbf{X}_p$ in the tangent space T_pM the number that is the rate of change at p of f along a curve with velocity vector $\mathbf{X}_p$. This was shown to be linear in Problem **4.8**.

We can define addition and scalar multiplication for linear functionals on V as follows: If α and β are linear functionals on V, and k is an element in K, then

$$(\alpha + \beta)(\mathbf{v}) \equiv \alpha(\mathbf{v}) + \beta(\mathbf{v}) \text{ and } (k\alpha)(\mathbf{v}) \equiv k(\alpha(\mathbf{v})).$$

With these operations the space of all linear functionals on V forms a vector space called the ***conjugate space*** (or ***dual space***) to V and is often denoted by V'. If $\{\mathbf{e}_1,\mathbf{e}_2,\ldots,\mathbf{e}_n\}$ is a basis for V, then define $\mathbf{e}^i$ to be the linear functional that assigns to each vector $\mathbf{v}$ its i-th coordinate with respect to the basis. Then for any linear functional α, we have

$$\begin{aligned}\alpha(\mathbf{v}) &= \alpha(\mathrm{v}^1\mathbf{e}_1 + \mathrm{v}^2\mathbf{e}_2 + \ldots + \mathrm{v}^n\mathbf{e}_n) = \\ &= \alpha(\mathrm{v}^1\mathbf{e}_1) + \alpha(\mathrm{v}^2\mathbf{e}_2) + \ldots + \alpha(\mathrm{v}^n\mathbf{e}_n) = \\ &= \mathrm{v}^1\alpha(\mathbf{e}_1) + \mathrm{v}^2\alpha(\mathbf{e}_2) + \ldots + \mathrm{v}^n\alpha(\mathbf{e}_n).\end{aligned}$$

Thus, if we define $\alpha(\mathbf{e}_i) \equiv \alpha_i$, then

$$\alpha(\mathbf{v}) = \Sigma \mathrm{v}^i\alpha_i = \Sigma\alpha_i\mathbf{e}^i(\mathbf{v}) = \Sigma(\alpha_i\mathbf{e}^i)(\mathbf{v}),$$

and we can write

$$\alpha = \alpha_1\mathbf{e}^1 + \alpha_2\mathbf{e}^2 + \ldots + \alpha_n\mathbf{e}^n.$$

Thus $\{\mathbf{e}^1,\mathbf{e}^2,\ldots,\mathbf{e}^n\}$ is a basis for V', and hence, the conjugate space has the same dimension as V.

Linear functionals are often called ***covectors*** because they are dual in the sense that if you apply a covector (linear functional) to a vector, you get a number (element of the field K) and, conversely, if you apply a vector to a covector, you also get a number. In fact, vectors in V can be expressed as linear functionals on the conjugate space V' by the identification:

$$\mathbf{v}(\alpha) \equiv \alpha(\mathbf{v}).$$

We express this by saying that $V = V''$.

A ***tensor of type*** (p,q) on the vector space V is a real-valued function F of p vectors variables and q covector (linear functional) variables:

$$F(\mathbf{v}_1,\mathbf{v}_2,\ldots,\mathbf{v}_p;\alpha_1,\alpha_2,\ldots,\alpha_q) \in \mathbb{R},$$

which is linear in each variable separately.

Examples of tensors:

1. Any linear functional is a tensor of type (1,0). Thus, if f is any differentiable real-valued function defined on a smooth surface M, then $\mathbf{X}_p \to \mathbf{X}_p f$ is a tensor of type (1,0) that is determined by the two numbers $\mathbf{x}_1 f$, $\mathbf{x}_2 f$.

2. Any vector $\mathbf{v}$ in V determines a unique tensor $\hat{\mathbf{v}}$ of type (0,1) by the identification: $\hat{\mathbf{v}}(\alpha) \equiv \alpha(\mathbf{v})$. Thus, it is possible to think of a vector as a tensor of type (0,1).

3. The Riemannian metric is a tensor of type (2,0) on the tangent space at a point.

4. A linear operator T determines a unique tensor $\mathrm{T}^\wedge$ of type (1,1) by the identification: $\mathrm{T}^\wedge(\mathbf{v},\alpha) = \alpha(\mathrm{T}(\mathbf{v}))$.

5. By Problem **8.5**, the Riemann curvature tensors are tensors of types (1,3) and (0,4).

But not everything is a tensor. For example, if we have a C^2 coordinate patch on a manifold M, then

$$\mathbf{X},\mathbf{Y} \to \nabla_{\mathbf{X}}\mathbf{Y}$$

is a vector-valued function, and thus its i-th coordinate $\mathbf{e}^i(\nabla_{\mathbf{X}}\mathbf{Y})$ is a number and thus,

$$\mathbf{X},\mathbf{Y},\alpha \to \alpha(\nabla_{\mathbf{X}}\mathbf{Y}) = \Sigma\alpha_i\mathbf{e}^i(\nabla_{\mathbf{X}}\mathbf{Y})$$

looks like it might be a tensor of type (2,1) but fails to be a tensor, because it depends on the values of the vector *field* $\mathbf{Y}$ near p and not just on $\mathbf{Y}_p$.

Appendix B

Analysis from a Geometric Point of View

B.1. Smooth Functions

We start with the notion of field of view as in Chapter 2, Problems **2.1** and **2.2**.

DEFINITION. A function f from a neighborhood U of a point p in $\mathbb{R}^n$ to $\mathbb{R}^m$ is ***smooth*** if the graph G of f is a smooth n-submanifold of $\mathbb{R}^n \times \mathbb{R}^m = \mathbb{R}^{n+m}$ and the projection of each tangent space of G to $\mathbb{R}^n$ is a one-to-one and onto. An ***n-submanifold*** M is a subset of a Euclidean space such that M is ***infinitesimally n-spatial***; that is, for every point p in M, there is an n-hyperplane T_p (called the ***tangent space*** at p) such that, for every tolerance $\tau = (1/N)$, there is a radius $\rho = (1/M)$, such that in any f.o.v. centered at p with radius less than ρ, the projection of M onto T_p is one-to-one and onto and moves each point less than $\tau\rho$ (we describe this by saying that if you zoom in on p, then M and T_p become indistinguishable). The submanifold is said to be ***smooth*** if the zooming is uniform in the sense that (for each tolerance) the same ρ can be used for every point in some neighborhood of p.

LEMMA. *The last sentence above is equivalent to saying that the tangent spaces vary continuously over M.*

The proof is essentially the same as Problems **2.2.e** and **3.1.e**.

Definition. If f is a smooth function from a neighborhood U in $\mathbb{R}^n$ to $\mathbb{R}^m$, then for each p in U, the ***differential***, df_p, is the linear function from $\mathbb{R}^n$ to $\mathbb{R}^m$ such that the tangent space T_p is the graph of the affine linear function

$$t(q) = f(p) + df_p(q - p).$$

In the terminology of Appendix **A.1**, we can more accurately say that df_p is a linear transformation from the tangent space $(\mathbb{R}^n)_p$ to the tangent space $(\mathbb{R}^m)_{f(p)}$.

Theorem B.1. *A function, which maps a neighborhood U of p in $\mathbb{R}^n$ to $\mathbb{R}^m$, is smooth (in the above geometric sense) if and only if it is* C^1 *(in the sense of having for every point p in U a differential df_p that varies continuously with p).*

The proof is essentially the same as the proofs of Problem **2.2.b,c,e** and Problem **3.1.e**.

B.2. Invariance of Domain

In the next section we will need the following result:

Theorem B.2. *Any continuous function that maps an open subset of n-space one-to-one to n-space is open (that is, the image of every open set is open).*

This result is commonly known as ***Brouwer's Invariance of Domain***. It was first proved in about 1910 by L.E.J. Brouwer. The proofs of this theorem involve the topological fields of dimension theory or homology theory, and all require a fair amount of machinery. There are proofs in any of the three books listed in the Bibliography in Section **Tp. Topology**. In the context of differentiable functions, there is an easier proof, which involves explicitly constructing a continuous inverse (see [**An:** Strichartz], the proof of Theorem 13.1.1.)

B.3. Inverse Function Theorem

Theorem B.3. *If f is a smooth function from n-space to n-space such that, for the point $p_o = (y_o, f(y_o))$ on the graph of f, the tangent space T_p projects one-to-one onto the* *range, then there is a neighborhood U of $x_0 = f(y_0)$ and a smooth function g from U to n-space such that $f(g(x)) = x$, for every x in U. Furthermore, g maps U one-to-one onto a neighborhood V of y_0 and $g(f(y)) = y$, for every y in V.*

Proof: This proof uses the Invariance of Domain but is otherwise shorter and more geometric than the usual proofs in analysis. The only nontrivial things to show are (a) that f is **one-to-one** in a neighborhood of y_0, and (b) that f maps a neighborhood of y_0 **onto** a neighborhood of x_0.

(a) Suppose f is not one-to-one in a neighborhood of y_0. Then there is a sequence of point pairs $\{a_n, b_n\}$ such that $f(a_n) = f(b_n)$, for all n. Let l_n be the line segment joining a_n to b_n. Applying the Mean Value Theorem for Space Curves (Problem **4.2.b**), there is a point c_n on l_n between a_n and b_n such that a vector tangent to the graph of $\lambda|l_n$ (and, therefore, tangent also to the graph of f) projects to a point on the range n-space. But then the tangent spaces to the graph of f cannot be varying continuously.

(b) The fact that f is onto a neighborhood follows from Invariance of Domain (**B.3**).

For an analytic proof of **B.2**, see [**An:** Strichartz], Theorem 13.1.2.

B.4. Implicit Function Theorem

THEOREM **B.4.1.** *Let $F(x,y)$ be a smooth function defined in a neighborhood of x_0 in $\mathbb{R}^n$ and y_0 in $\mathbb{R}^m$ taking values in $\mathbb{R}^m$, with $F(x_0,y_0) = c$. Then, if the function $f(x) = F(x_0,y)$ is such that, for the point $p_0 = (x_0,y_0,f(y_0))$ on the graph of f, the tangent space T_p projects one-to-one onto the range, then there is a neighborhood U of x_0 and a smooth function h from U to $\mathbb{R}^m$ such that $h(x_0) = y_0$ and $F(x,h(x)) = c$ for every x in U.*

Note that the condition on the graph of f is equivalent to the analytic condition that $F_y(x_0,y_0)$ is invertible, where F_y is the submatrix of dF corresponding to using only the partial derivatives with respect to y.

Proof: We will describe three different proofs:

1. Define $f(x,y) = (x,F(x,y))$. Then it is easy to check that f satisfies the hypotheses of Theorem **B.3**. Thus, if there is a smooth inverse function g defined on a neighborhood of $(x_0,F(x_0,y_0))$, then there is a function $h(x)$ such that $g(x,c) = (x,h(x))$. This h is the desired function.

2. It is possible to construct a direct geometric proof (using Invariance of Domain) along the same lines as the proof of Theorem **B.3**.
3. There is an analysis proof that explicitly constructs the function h. (See [**An:** Strichartz], Theorem 13.1.1.)

THEOREM B.4.2. *Let* $F : \mathbb{R}^n \rightarrow \mathbb{R}^{n-m}$ *be a* C^1 *function, and suppose* $dF(x)$ *has maximal rank* $n - m$ *at every point on a level set*

$$M = \{ x \mid F(x) = c \}.$$

Then M *is a* C^1 *m-submanifold of* $\mathbb{R}^n$.

We can prove this as a corollary of **B.4.1**, (see [**An:** Strichartz], Theorem 13.2.2.) But there is a more geometric proof. First, change the hypotheses to geometric ones. That dF has maximal rank at p is equivalent to the tangent space $T_{p,F(x)}$ of the graph of F projecting *onto* the range space. Now, if we take the inverse of c under this projection, we get a linear m-dimensional subspace of the tangent space. The projection of this m-subspace onto the domain is a tangent space of the level set M. Thus, we have proved:

Theorem B.4.3. *Let* $F : \mathbb{R}^{n+m} \rightarrow \mathbb{R}^m$ *be a smooth function, such that, for every point* $p = (x,y,c)$ *on the graph of the level set*

$$M = \{ x \mid F(x,y) = c \},$$

the tangent space T_p *projects onto the range. Then* M *is a* C^1 *m-submanifold of* $\mathbb{R}^n$.

Appendix C

Computer Scripts

Those readers who have access to computer systems running Maple©, Mathematica©, Derive©, or similar software can use these systems to facilitate gaining geometric intuition and imagination of the concepts of differential geometry. However, the current state-of-the-art technology for generally available computer graphing programs is not capable of producing what would be the most useful displays. For example, it is not currently possible, with widely useable programs, to view a curve on a surface and to use the mouse to dynamically move a point along the curve and see displayed the three curvature vectors—intrinsic (geodesic), extrinsic, and normal.

In this appendix we have included several computer exercises for Maple, and these and additional scripts are also available for downloading on-line at

```
ftp://math.cornell.edu/pub/Henderson/diff_geom.
```

The first section contains a file with the definitions of several functions used by the other scripts. The other scripts are labeled according to the problem in the text to which they are most connected.

Standard Functions

These are definitions of commonly used functions that can be called by other scripts, and referred in the other scripts as the file with name `diffgeo_defs`. If you save it as a file with a different name, then you will have to change the appropriate lines in the scripts.

```
with(plots):
with(linalg):
macro(medgreen = COLOR(RGB,0,0.8,0.5));
macro(medgrey = COLOR(RGB,0.6,0.6,0.6));
```

```
vecnorm:= (vec) -> sqrt(dotprod(vec,vec)):
vecnormalize:= (vec)->evalm((1/vecnorm(vec))*vec):
Tangent := proc(t,func)
   local Velocity, VelEval, VelLength;
      Velocity := map(diff, func(x), x);
      VelEval := map(evalf,(subs(x=t, Velocity)));
      VelLength := sqrt(dotprod(VelEval,VelEval));
   evalm((1/VelLength)*VelEval)
end:
Tanvec := proc(t,func)
   polygonplot3d([func(t),convert(evalm(func(t)+
      Tangent(t,func)), list)],color=medgreen)
end:
CNormal := proc(t,func)
   local TanPrime, TanPrimeEval,
      TanPrimeLength;
      TanPrime:= map(diff,Tangent(x,func),x);
      TanPrimeEval:=
         map(evalf,subs(x=t,eval(TanPrime)));
      TanPrimeLength:= sqrt(dotprod(TanPrimeEval,
      TanPrimeEval));
   evalm((1/TanPrimeLength)*TanPrimeEval)
end:
Normvec := proc(t,func)
   polygonplot3d([func(t),convert(evalm(func(t)+
      CNormal(t,func)),list)],color=red)
end:
Binormal := (t,func) ->
   crossprod(Tangent(t,func),CNormal(t,func)):
BiVec := proc(t,func)
   polygonplot3d([func(t),convert(evalm(func(t)+
      Binormal(t,func)), list)],color=blue)
end:
Velocity := proc(t, func)
   local fprime;
      fprime := map(diff, func(x), x);
   map(evalf, subs(x=t, fprime))
end:
sprime := proc(t, func)
```

```
   sqrt(dotprod(Velocity(t,func),Velocity(t,func)))
end:
VCurvature := proc(t, func)
   local TanPrime, TanPrimeEval;
      TanPrime := map(diff,Tangent(x,func),x);
      TanPrimeEval := map(evalf,subs(x=t,
         eval(TanPrime)));
   evalm((1/sprime(t,func))*TanPrimeEval)
end:
scurvature := proc(t,func)
   sqrt(dotprod(VCurvature(t,func),
         VCurvature(t,func)))
end:
CurvVec := proc(t,func)
   polygonplot3d([func(t),convert(evalm(func(t)+
      VCurvature(t,func)), list)], color=magenta)
end:
CurvVec2 := proc(t,func)
   polygonplot3d([func(t),convert(evalm(func(t)+
      VCurvature(t,func)), list)],
      color=magenta, thickness=2)
end:
GaussMap:= proc(func)
   local func1, func2;
      func1 := map(diff, func(x1,x2), x1);
      func2 := map(diff, func(x1,x2), x2);
   vecnormalize(crossprod(func1,func2))
end:
GMapEval := proc(func,u,v)
   evalf(subs(x1=u,x2=v, GaussMap(func)))
end:
FirstFundFormDet := proc(func)
   local func1, func2, g11, g12, g22;
      func1 := map(diff, func(x1,x2), x1);
      func2 := map(diff, func(x1,x2), x2);
      g11 := dotprod(func1,func1);
      g12 := dotprod(func1,func2);
      g22 := dotprod(func2,func2);
   g11*g22 - g12^2
```

```
end:
SurfaceNorm := proc(func)
   local func1, func2;
      func1 := map(diff, func(x1,x2), x1);
      func2 := map(diff, func(x1,x2), x2);
   vecnormalize(crossprod(func1,func2))
end:
SecondFundFormDet := proc(func)
   local func11, func12, func22, L11, L12,L22;
      func11 := map(diff, func(x1,x2), x1, x1);
      func12 := map(diff, func(x1,x2), x1, x2);
      func22 := map(diff, func(x1,x2), x2, x2);
      L11 := dotprod(func11, SurfaceNorm(func,
         x1, x2), orthogonal);
      L12 := dotprod(func12, SurfaceNorm(func,
         x1, x2), orthogonal);
      L22 := dotprod(func22, SurfaceNorm(func,
         x1, x2), orthogonal);
   L11*L22 - L12^2
end:
GaussCurv := proc(func,u,v)
   evalf(subs(x1=u, x2=v, SecondFundFormDet(func)/
      FirstFundFormDet(func)))
end:
```

Computer Exercise 1.6: Strake

In this exercise, you can vary the inner radius of a strake.

```
> with(plots):
> setoptions3d(scaling=constrained);
```

First, input a radius between 0 and 2:

```
> r := 1;
```

Create a cylinder with that radius:

```
> cylinder := plot3d([r*cos(u), r*sin(u), v],
     u=0..2*Pi, v=0..2, color=green):
```

Create a strake with inner radius `r` and outer radius 2:

```
> strake := plot3d([((1-t)*r+ 2*t)*cos(u),
```

```
((1-t)*r +2*t)*sin(u), u/Pi], u=0..2*Pi,
t=0..1, color=blue):
```

Display them together:

```
> display({strake, cylinder});
```

We suggest viewing the surfaces in the style "`patch w/o grid`" with one of the lighting schemes. Try various values for `r`, including `r=0` and `r=2`. When you are finished with this exercise, hit the enter key after the following line:

```
> r := 'r';
```

This will make `r` a variable again. If you don't do this, the computer will continue to think that `r=1` (or whatever your last radius value was).

Computer Exercise 1.7: Surfaces of Revolution

In this exercise, we will take a plane curve and turn it into a surface of revolution.

```
> with(plots):
> setoptions3d(scaling=constrained);
```

First, the plane curve:

```
> f:= (x) -> sin(x) + 2;
```

Now we create the surface of revolution, defined by (u,v) goes to (u, `f`(u)cos(v), `f`(u)sin(v)):

```
> plot3d([u, f(u)*cos(v), f(u)*sin(v)], u=-Pi..Pi,
      v=0..2*Pi, axes=normal, orientation=[90,90],
      color=blue);
```

Notice that when the surface first appears, we are looking down the y-axis. The upper boundary of the surface is `f(x)` and the lower boundary is `-f(x)`. Try rotating the surface to see how it looks from various angles. Experiment with other plane curves by changing the definition of `f` on your worksheet. Note that you may want to vary the range of u in the "`plot3d`" line. Can you make a cylinder? a cone? a sphere? a paraboloid? Keep these surfaces in mind as you do Problem 1.7.

Computer Exercise 1.9: Surface as Graph of a Function

This exercise allows you to test out your answers to Problem 1.9.

```
> with(plots):
> setoptions3d(scaling=constrained):
```

First, input a function of `x` and `y` (the default surface is a monkey saddle):

```
> f:= (x,y) -> (1/2)*x*(x^2 - 3*y^2);
```

Then display it:

```
> plot3d(f(x,y),x=-1..1,y=-1..1,color=blue);
```

We suggest using the style "`patch`" and one of the lighting schemes. Remember that when you change your surface equation, you may want to vary the bounds on x and y in the "`plot3d`" statement.

Computer Exercise 2.2: Tangent Vectors to Curves

In this exercise, you can input a space curve and see the **tangent vectors** at various points.

```
> with(plots):
> with(linalg):
> setoptions3d(scaling=constrained);
> read(`diffgeo_defs`);
```

First, the space curve:

```
> f:= (x) -> [cos(x), sin(x), x/7];
```

Now we define the domain (from `a` to `b`) and the number of divisions of the domain (`div`). Eventually, we will find the tangent vector at the end-points (`points`) of each division.

```
> a:=0;
> b:=4*Pi;
> div:=10;
> points := {seq( a + i*(b-a)/div, i=0..div)}:
```

Now we get the tangent vector by differentiating `f` to get the velocity and then normalizing.

```
> Tangent(t,f);
```

Here we create the vectors that will appear on the screen.

```
> Vectors := map(Tanvec, points, f):
```

Here we calculate the coordinates we need to put the curve on the screen.

```
> Curve := spacecurve(f(x),x=a..b,thickness=2,
      shading=zgrayscale):
```

And finally, we display the curve and its tangent vectors.

```
> display({Curve} union Vectors);
```

Now try other curves by changing `f` on your worksheet. Remember you may also want to change your domain and/or the number of divisions.

Computer Exercise 2.3: Curvature and Tangent Vectors

In this exercise, you can input a curve and see the **curvature and tangent vectors** at various points.

```
> with(plots):
> with(linalg):
> read(`diffgeo_defs`):
> setoptions3d(scaling=constrained):
```

First, the space curve:

```
> f := (x) -> [cos(x), sin(x), x/7];
```

Now we define the domain (from `a` to `b`) and the number of divisions of the domain (`div`). Eventually, we will find the tangent vector at the endpoints (`points`) of each division.

```
> a:=0;
> b:=4*Pi;
> div:=10;
> points := {seq(a + i*(b-a)/div, i=0..div)}:
```

Here we create the vectors that will appear on the screen.

```
> TVectors := map(Tanvec, points, f):
> CVectors := map(CurvVec, points,f):
```

Here we calculate the coordinates we need to put the curve on the screen.

```
> Curve := spacecurve(f(x),x=a..b,thickness=2,
      shading=zgrayscale):
```

And finally, we display the curve and its tangent vectors.

```
> display({Curve} union TVectors union CVectors);
```

The default curve is a helix. Experiment with different helices (try different values of c and d in the line

```
> f := (x) -> [cos(x), sin(x), x/7];
```

and see how the curvature vector changes). Try the logarithmic spiral as well–vary its domain to see different positions of the curve. What do you observe?

Computer Exercise 2.4a: Osculating Planes

In this exercise, you can input a space curve and see the **osculating plane**, tangent vector, and normal vector at various points.

```
> with(plots):
> with(linalg):
> setoptions3d( scaling=constrained );
> read(`diffgeo_defs`):
```

The default curve is a helix. First, input the curve:

```
> f:= (x) -> [2*cos(x), 2*sin(x), x/5];
```

Now define the curve domain (from a to b) and the number of divisions in the domain (div). Eventually, we will find the osculating plane at the endpoints of each division.

```
> a:=0;
> b:=4*Pi;
> div:=5;
> points := {seq( a + i*(b-a)/div, i=0..div)}:
```

Here we calculate the coordinates we need to put the curve on the screen.

```
> Curve := spacecurve(f(x), x=a..b, thickness=2,
      shading=zgrayscale):
```

Now we define the osculating plane, which is spanned by the tangent vector and the normal vector.

```
> OscPlane := proc(t)
   plot3d(evalm(x*Tangent(t,f)+ y*CNormal(t,f)+
   f(t)), x=-1/2..1/2, y=-1/2..1/2, color=yellow,
   numpoints=81)
  end:
```

Here we create the vectors and planes that will appear on the screen.

```
> TVectors := map(Tanvec, points, f):
> NVectors := map(Normvec, points, f):
> Planes := map(OscPlane, points):
```

And now we display them.

```
> display({Curve} union TVectors union NVectors
   union Planes);
```

Computer Exercise 2.4b: Osculating Circles

In this exercise, you can input a space curve and see the **osculating circle**, tangent, and curvature vectors at a given point.

```
> with(plots):
> with(linalg):
> macro(purple=COLOR(RGB, .5,0,.5));
> read(`diffgeo_defs`):
> setoptions3d(scaling=constrained);
```

First, the space curve:

```
> f:= (x) -> [1.5*cos(x), 1.5*sin(x), x/3];
```

Now we define the domain (from `a` to `b`) and pick the point where we'll find the osculating circle by choosing a point in the domain.

```
> a:=0;
> b:=4*Pi;
> Point := Pi/3;
```

Here we calculate the coordinates we need to put the curve on the screen.

```
> Curve := spacecurve(f(x), x=a..b, thickness=2,
   shading=zgrayscale):
```

Now we calculate the coordinates of the tangent and curvature vectors so that we can put them on the screen.

```
> TVector := Tanvec(Point,f):
> CVector := CurvVec(Point,f):
```

Now we find the osculating circle. Note that in the plane, the circle of radius `1/scurvature` centered at `[0, 1/scurvature]` is given by

```
[(1/scurvature)*cos(x),(1/scurvature)*(1+sin(x))]
```

(Here **scurvature** is the scalar curvature. **VCurvature** is the curvature vector.) To move this circle to the osculating plane, we map [1,0] to the tangent vector and [0,1] to **(1/scurvature))*VCurvature**.

```
> OscCircle := proc(t,func,x)
    evalm((1/scurvature(t,func)) * cos(x) *
    Tangent(t,func) +
    ((1/scurvature(t,func))^2) * (1 + sin(x)) *
    VCurvature(t,func) + func(t))
  end:
```

Now we compute the coordinates we need to put the circle on the screen:

```
> Circle := spacecurve(OscCircle(Point, f, x),
    x=0..2*Pi, color=purple):
```

Finally, we display everything.

```
> display({Curve, Circle, TVector, CVector});
```

Computer Exercise 2.6: Frenét Frame

In this exercise, you can input a space curve and see the Frenét frame at various points.

```
> with(plots):
> with(linalg):
> setoptions3d(scaling=constrained);
> read(`diffgeo_defs`):
```

First, the space curve:

```
> f:= (x) -> [2*cos(x), 2*sin(x), x/5];
```

Now we define the domain (from a to b) and the number of divisions of the domain (**div**). Eventually, we will find the Frenét frame at the endpoints (**points**) of each division.

```
> a:=0;
> b:=4*Pi;
> div:=10;
> points := {seq( a + i*(b-a)/div, i=0..div)}:
```

Here we calculate the coordinates we need to put the curve on the screen.

```
> Curve := spacecurve(f(x), x=a..b, thickness=2,
    shading=zgrayscale):
```

Here we create the vectors that will appear on the screen. The tangent vectors are green, the normals are red, and the binormals are blue. For the definitions of the functions **Tanvec**, **Normvec**, and **BiVec**, please see the included file.

```
> TVectors := map(Tanvec, points, f):
> NVectors := map(Normvec, points, f):
> BVectors := map(BiVec, points,f):
```

Finally, we put everything on the screen.

```
> display({Curve} union TVectors union NVectors
   union BVectors);
```

Computer Exercise 3.1: Tangent Planes to Surfaces

In this exercise, you can input a surface and a point and see the tangent plane at that point. The default surface is a torus.

```
> with(plots):
> with(linalg):
> read(`diffgeo_defs`):
```

First, the surface domain (`x` goes from `a1` to `a2`, `y` from `b1` to `b2`) and a specific point in that domain **(pointx,pointy)**:

```
> a1 := 0;
> a2 := 2*Pi;
> b1 := 0;
> b2 := 2*Pi;
> pointx := Pi/3;
> pointy := 3*Pi/4;
```

The surface:

```
> f := (x,y) -> [(cos(x) + 2)*cos(y),(cos(x) +
   2)*sin(y), sin(x)];
> Surface := plot3d(f(x,y), x=a1..a2, y=b1..b2,
   color-bluc):
```

We find the tangent plane by finding the partial derivative of `f` at the chosen point.

```
> XDeriv := map(diff, f(x,y), x);
> XDerivEval := evalf(subs(x=pointx,
   y=pointy,XDeriv)):
> YDeriv := map(diff, f(x,y),y);
```

```
> YDerivEval := evalf(subs(x=pointx, y=pointy,
   YDeriv)):
> Plane := plot3d(convert(evalm(x*XDerivEval +
   y*YDerivEval + f(pointx,pointy)), list),
   x=-1..1, y=-1..1, color=green, numpoints=100):
```

Find the coordinate curves :

```
> XCurveEq := (t) -> f(t,pointy);
> XCurve := spacecurve(XCurveEq(t), t=a1..a2,
   color=yellow, thickness=2):
> YCurveEq := (t) -> f(pointx,t);
> YCurve := spacecurve(YCurveEq(t), t=b1..b2,
   color=orange, thickness=2):
```

Putting the vectors on the screen:

```
> XVector := polygonplot3d([f(pointx,pointy),
   convert(evalm(f(pointx,pointy)+ XDerivEval),
   list)], color=yellow, thickness=2):
> YVector := polygonplot3d([f(pointx,pointy),
   convert(evalm(f(pointx,pointy)+ YDerivEval),
   list)], color=orange, thickness=2):
```

Display everything:

```
> display({Surface, Plane, XVector, YVector,
   XCurve, YCurve});
```

Computer Exercise 3.2a: Curves on a Surface

In this exercise, you can input a surface, its domain, and a curve in its domain, and see the image of the curve on the surface. The default surface is a torus.

```
> with(plots):
> setoptions3d(scaling=constrained);
```

First, the domain (`x` goes from `a1` to `a2`, `y` from `b1` to `b2`):

```
> a1:= 0;
> a2:= 2*Pi;
> b1:= 0;
> b2:= 2*Pi;
> domain:= plot3d([x,y,0], x=a1..a2, y=b1..b2,
   color=blue, style=wireframe):
```

The curve in the domain (that is, a plane curve):

```
> c1 := 0;
> c2 := 2*Pi;
```

The domain of the curve is from `c1` to `c2`.

```
> f1 := (x) -> x;
> f2 := (x) -> Pi;
> DomCurve := spacecurve([f1(x),f2(x),0], x=c1..c2,
   color=yellow, thickness=2):
```

The surface:

```
> g := (x,y) -> [(cos(x) + 2)*cos(y),(cos(x) +
   2)*sin(y), sin(x)];
> Surface := plot3d(g(x,y),x=a1..a2,y=b1..b2,
   color=blue):
```

And the curve on the surface:

```
> SurfCurveEq:= (x)->subs(u=f1(x), v=f2(x),
   g(u,v));
> SurfCurve:= spacecurve(SurfCurveEq(x), x=c1..c2,
   color=yellow,thickness=2):
```

Now display everything. You will get two separate windows. We recommend viewing the surface in the "`patch`" style, possibly with one of the lighting schemes.

```
> display({domain, DomCurve});
> display({Surface, SurfCurve});
```

You've seen what happens to the line `y=Pi` when it is mapped to the torus. What happens to other straight lines in the domain of the torus? Are their images intrinsically straight in the torus?

Computer Exercise 3.2b: Extrinsic Curvature Vectors

In this exercise, you can input a surface, its domain, and a curve in its domain. You'll see the image of the curve on the surface and the extrinsic curvature vector of this new curve. This exercise is primarily a tool for exploring Problem **3.2**.

```
> with(plots):
> setoptions3d(scaling=constrained);
> read(`diffgeo_defs`):
```

First, the domain (x goes from a1 to a2, y goes from b1 to b2):

```
> a1:= 0;
> a2:= 2*Pi;
> b1:= 0;
> b2:= 4;
> domain:= plot3d([x,y,0], x=a1..a2, y=b1..b2,
   color=blue, style=wireframe):
```

The curve in the domain (that is, a plane curve):

```
> c1 := 0;
```

The domain of the plane curve goes from c1 to c2.

```
> c2 := 2*Pi;
> Point := Pi;
> f1:= (x) -> x;
> f2:= (x) -> 0.5*x;
> DomCurve:= spacecurve([f1(x),f2(x),0], x=c1..c2,
   color=yellow, thickness=2):
```

The surface:

```
> g := (x,y) -> [cos(x), sin(x), y];
> Surface := plot3d(g(x,y), x=a1..a2, y=b1..b2,
   color=blue):
```

And the curve on the surface:

```
> SurfCurveEq := (x) -> subs(u=f1(x), v=f2(x),
   g(u,v));
> SurfCurve:= spacecurve(SurfCurveEq(x),  x=a1..a2,
   color=yellow,thickness=2):
```

Now we find the extrinsic curvature vector of the surface curve.

```
> CVector := CurvVec2(Point, SurfCurveEq):
```

Now display everything. You will get two separate windows. Note: We recommend that you look at the surface in the wireframe style. This will make it easier to see where the curvature vector is.

```
> display({domain, DomCurve});
> display({Surface, SurfCurve, CVector});
```

You may also want to restrict the domain of your surface (for example, look at only half a cylinder), again to make it easier to see the curvature vector.

Computer Exercise 3.3: The Three Curvature Vectors

In this exercise, you can input a surface, a curve on that surface, and a point, and see the extrinsic, normal, and geodesic curvature vectors at that point. The default surface is half a cylinder.

```
> with(plots):
> setoptions3d(scaling=constrained);
> read(`diffgeo_defs`):
```

First, the domain (x goes from `a1` to `a2`, y goes from `b1` to `b2`):

```
> a1:= 0;
> a2:= Pi;
> b1:= 0;
> b2:= 4;
> domain:= plot3d([x,y,0], x=a1..a2, y=b1..b2,
   color=blue, style=wireframe):
```

The curve in the domain (that is, a plane curve): The domain of the plane curve goes from `c1` to `c2`. **`Point`** is the point at which we'll show the curvature vectors.

```
> c1 := 1;
> c2 := 3;
> Point := 2;
> f1:= (x) -> x;
> f2:= (x) -> 1 + (x-2)^2;
> domcurve := spacecurve([f1(x), f2(x),0],
   x=c1..c2, color=yellow, thickness=2):
```

The surface:

```
> g := (x,y) -> [cos(x), sin(x), y];
> surf:= plot3d(g(x,y), x=a1..a2, y=b1..b2,
   color=blue):
```

And the curve on the surface:

```
> SurfCurveEq := (x) -> subs(u=f1(x), v=f2(x),
   g(u,v));
> SurfCurve:= spacecurve(SurfCurveEq(x), x=c1..c2,
   color=yellow,thickness=2):
```

Now we find the extrinsic curvature vector of the surface curve and do the calculations we need to put it on the screen.

```
> ExtVector := CurvVec2(Point, SurfCurveEq):
> ExtCurv := VCurvature(Point, SurfCurvEq):
```

Now we break the curvature vector into its normal and geodesic components. First, find the surface normal and get the tangent plane:

```
> XDeriv := map(diff, g(x,y), x);
> XDerivEval := evalf(subs(x=f1(Point),
   y=f2(Point), XDeriv)):
> YDeriv := map(diff, g(x,y),y);
> YDerivEval := evalf(subs(x=f1(Point),
   y=f2(Point), YDeriv)):
> SurfaceNormal := vecnormalize(crossprod
   (XDerivEval, YDerivEval)):
> Plane := plot3d(convert(evalm(x*XDerivEval +
   y*YDerivEval + g(f1(Point), f2(Point))), list),
   x=-1..1, y=-1..1, color=green, numpoints=64):
```

Now project the extrinsic curvature vector onto the surface normal to get the normal curvature vector. The geodesic curvature vector is the projection of the extrinsic curvature vector onto the tangent plane, but using the fact that (extrinsic curvature vector = normal curvature vector + geodesic curvature vector) saves computation time.

```
> NormCurv := evalm((dotprod
   (ExtCurv,SurfaceNormal))*SurfaceNormal):
> GeoCurv := evalm(ExtCurv - NormCurv):
> NormVector := polygonplot3d ([g(f1(Point),
   f2(Point)), convert(add(g(f1(Point), f2(Point)),
   NormCurv), list)], color=black, thickness=2):
> GeoVector := polygonplot3d ([g(f1(Point),
   f2(Point)), convert(add(g(f1(Point), f2(Point)),
   GeoCurv), list)], color=coral, thickness=2):
```

Now display everything. You will get two separate windows. Note: We recommend that you look at the surface in the wireframe style. This will make it easier to see where the curvature vectors are.

```
> display({domain, domcurve});
> display({surf, SurfCurve, GeoVector, NormVector,
   Plane, ExtVector});
```

Computer Exercise 3.4: Ruled Surfaces

In this exercise, we take a curve (`alpha`) and a unit vector (`r`) and create a ruled surface.

```
> with(plots):
> setoptions3d(scaling=constrained):
```

First, define the curve and its domain (from `a1` to `a2`):

```
> a1:=0;
> a2:=2*Pi;
> alpha := (t) -> [cos(t), sin(t), t/5];
```

Now define the vector:

```
> r := (t) -> [0, cos(t), sin(t)];
```

Now define the surface and the domain of `s` (from `b1` to `b2`). `s` multiplies the vector `r` and determines the width of the surface.

```
> b1:= -1/2;
> b2:= 1/2;
> Surface := (t,s) -> evalm(alpha(t)+ s*r(t));
```

Now choose the number of divisions (`div`) of the curve domain. We'll draw the rulings (calculated by `Rule`) at the endpoints (`points`) of each division.

```
> div :=10;
> points := {seq(a1+i*(a2-a1)/div, i=0..div)}:
> Rule := proc(t)
   polygonplot3d([evalm(alpha(t) + b1*r(t)),
   evalm(alpha(t) + b2*r(t))], color=navy,
   thickness=2)
  end:
```

Now we calculate the coordinates we need to put everything on the screen.

```
> RuledSurf := plot3d(Surface(t,s), t=a1..a2,
   s=b1..b2, style=wireframe):
> Curve := spacecurve(alpha(t), t=a1..a2,
   color=navy, thickness=2):
> Rulings := map(Rule, points):
```

Finally, we display everything.

```
> display({RuledSurf, Curve} union Rulings);
```

Computer Exercise 5.2: Non-dissectable Polyhedron

This exercise will display on the screen the polyhedron pictured in Figure 5.4. We suggest using "Patch w/ contour." The programming here isn't especially important (or interesting). Feel free to just hit enter after prompt without reading too carefully—the point is to get the polyhedron on the screen and play with it.

```
> with(plots):
```

First, we define a function that takes three points and sides in the triangle they determine.

```
> Triangle := proc(vec1,vec2,vec3)
    evalm(x*(vec2-vec1)+ y*(1-x)*(vec3-vec1)+ vec1)
end:
> MakeFace := proc(vec)
    plot3d(Triangle(vec[1],vec[2],vec[3]),
        x=0..1, y=0..1, color=[vec[1][3],
        vec[2][3],vec[3][3]], numpoints=36)
end:
```

Now we list the vertices that determine each face of the polyhedron.

```
> a := Pi/12;
> h:=1;
> PreFaces :=
    {[[1,0,0], [-1/2,sqrt(3)/2,0],
    [-1/2,-sqrt(3)/2,0]], [[cos(a), sin(a),h],
    [cos(a+2*Pi/3), sin(a+2*Pi/3), h],
    [cos(a+4*Pi/3), sin(a+4*Pi/3), h]],
    [[1,0,0], [cos(a), sin(a), h], [-1/2,
    -sqrt(3)/2, 0]], [[1,0,0],[cos(a), sin(a), h],
    [cos(a+2*Pi/3), sin(a+2*Pi/3), h]], [[1,0,0],
    [-1/2, sqrt(3)/2, 0], [cos(a+2*Pi/3),
    sin(a+2*Pi/3), h]], [[cos(a), sin(a), h], [-1/2,
    -sqrt(3)/2,0], [cos(a+4*Pi/3), sin(a+4*Pi/3),
    h]], [[-1/2, sqrt(3)/2, 0], [-1/2, -sqrt(3)/2,
    0], [cos(a+4*Pi/3), sin(a+4*Pi/3), h]], [[-1/2,
    sqrt(3)/2,0], [cos(a+2*Pi/3), sin(a+2*Pi/3), h],
    [cos(a+4*Pi/3), sin(a+4*Pi/3), h]]}:
```

Now we make the actual faces out of these vertices and display the polyhedron.

```
> Faces := map(MakeFace, PreFaces):
> display(Faces);
```

Computer Exercise 5.5: Sign of (Gaussian) Curvature

In this exercise you can display a surface with areas of positive (Gaussian) curvature magenta and areas of negative (Gaussian) curvature blue. The formula used to compute the curvature is developed in Problem **7.1**.

```
> with(plots):
> with(linalg):
> read(`diffgeo_defs`):
> setoptions3d(scaling=constrained):
```

First, define the surface:

```
> f := (x,y) -> [(cos(x) + 2)*cos(y), (cos(x) + 2)
   *sin(y), sin(x)]:
```

Now display it, coloring areas of positive Gaussian curvature magenta and areas of negative Gaussian curvature blue.

```
> Surface := plot3d(f(x,y), x=0..2*Pi, y=0..2*Pi,
   color= [Heaviside(GaussCurv(f,x,y)), 0,1]):
> display(Surface);>
```

Computer Exercises 6.1: Multiple Principle Directions

This exercise goes with Problem **6.1.c** and will allow you to display the surfaces mentioned there. Remember that `f` must be twice differentiable and that we need `f(0)= 0 = f'(0)` and `f(x)= f(-x)`.

```
> with(plots);
> setoptions3d(scaling=constrained);
> f := (theta) -> 1 - cos(4*theta);>
   cylinderplot([r,theta, f(theta)*r^2], r=0..1/2,
   theta=0..2*Pi, color=orange);
```

You may want to restrict the domain of `theta` and look at only, say, half or three-quarters of your surface. We recommend using the patch style and one of the lighting schemes.

Computer Exercise 6.3: Gauss Map

In this exercise, you can input a surface and see the image of a portion of it under the Gauss map. The default surface is a torus.

```
> with(plots):
> with(linalg):
> setoptions3d(scaling=constrained):
> read(`diffgeo_defs`):
```

First, the domains. `Surfdomain` is the domain of the surface (`x` goes from `a1` to `a2`, `y` goes from `b1` to `b2`). `Gaussdomain` is a subset of `Surfdomain` (`x` goes from `c1` to `c2`, `y` goes from `d1` to `d2`). You will see the image of the `Gaussdomain` under the Gauss map. We will also outline the `Gaussdomain` to make it easier to see what is going on.

```
> a1 := 0;
> a2 := 2*Pi;
> b1 := 0;
> b2 := 2*Pi;
> c1 := 2*Pi/3;
> c2 := 4*Pi/3;
> d1 := 2*Pi/3;
> d2 := 4*Pi/3;
> surfdomain := plot3d([x,y,0], x=a1..a2, y=b1..b2,
   color=blue, numpoints=100):
> Gaussdomain := plot3d([x,y,0], x=c1..c2,
   y=d1..d2, color=yellow, numpoints=144):
> outline1 := spacecurve([x,d1,0], x=c1..c2,
   color=green, thickness=2):
> outline2 := spacecurve([c1,y,0], y=d1..d2,
   color=red, thickness=2):
> outline3 := spacecurve([x,d2,0], x=c1..c2,
   color=magenta, thickness=2):
> outline4 := spacecurve([c2,y,0], y=d1..d2,
   color=black, thickness=2):
```

The surface and the image of the `Gaussdomain` and its outline in the surface (`SurfPatch` and the `surflines`):

```
> f := (x,y) -> [(cos(x) +2)*cos(y), (cos(x) +2)
   *sin(y), sin(x)];
```

```
> Surface := plot3d(f(x,y), x=a1..a2, y=b1..b2,
   color=blue):
> SurfPatch := plot3d(f(x,y), x=c1..c2, y=d1..d2,
   color=yellow):
> surfline1 := spacecurve(f(x,d1), x=c1..c2,
   color=green, thickness=2, numpoints=12):
> surfline2 := spacecurve(f(c1,y), y=d1..d2,
   color=red, thickness=2, numpoints=12):
> surfline3 := spacecurve(f(x,d2), x=c1..c2,
   color=magenta, thickness=2, numpoints=12):
> surfline4 := spacecurve(f(c2,y), y=d1..d2,
   color=black, thickness=2, numpoints=12):
```

Now the sphere for the Gauss map:

```
> SphereEq := (x,y) -> [sin(x)*cos(y),
   sin(x)*sin(y), cos(x)];
> Sphere := plot3d(SphereEq(x,y), x=0..Pi,
   y=0..2*Pi, color=turquoise, numpoints=225):
```

The image of the `Gaussdomain` and its outline:

```
> GaussImage := plot3d(GMapEval(f,x,y), x=c1..c2,
   y=d1..d2, color=yellow, numpoints=225):
> Gline1 := spacecurve(GMapEval(f,x,d1), x=c1..c2,
   color=green, thickness=2,  numpoints=10):
> Gline2 := spacecurve(GMapEval(f,c1,y), y=d1..d2,
   color=red, thickness=2,  numpoints=10):
> Gline3 := spacecurve(GMapEval(f,x,d2), x=c1..c2,
   color=magenta, thickness=2, numpoints=10):
> Gline4 := spacecurve(GMapEval(f,c2,y), y=d1..d2,
   color=black, thickness=2, numpoints=10):
```

Now we display everything. You may want to look at the surface window and the `Gausswindow` in the wireframe style—it makes it easier to see the outline of the yellow patch.

```
> display({surfdomain, Gaussdomain, outline1,
   outline2, outline3, outline4});
> display({Surface, SurfPatch, surfline1,
   surfline2, surfline3, surfline4});
> display({Sphere, GaussImage, Gline1, Gline2,
   Gline3, Gline4});
```

Computer Exercise 6.6: Helicoid to Catenoid

In this exercise, you can look at a helicoid, a catenoid, and the intermediate surfaces.

```
> with(plots):
> setoptions3d(scaling=constrained):
```

`a` is a constant that affects the shape of the surfaces.

```
> a:=2;
```

Helicat is a function that interpolates between the helicoid

$$(x,y) \rightarrow [x\cos(ay),\ x\sin(ay), y]$$

and the catenoid

$$(x,y) \rightarrow \left[\tfrac{1}{a}\sqrt{1+(ax)^2}\cos(ay), \tfrac{1}{a}\sqrt{1+(ax)^2}\sin(ay), \tfrac{1}{a}arcsinh(ax)\right].$$

`Helicat(x,y,0)` is the helicoid, `Helicat(x,y,1)` is the catenoid, and `Helicat(x,y,t)`, where `t` is between `0` and `1` is an intermediate surface.

```
> Helicat := (x,y,t)->
    [((1/a)*(sqrt(1+(a*x)^2))*t+ (1-t)*x)*cos(a*y),
    ((1/a)*(sqrt(1+ (a*x)^2))*t+ (1-t)*x)*sin(a*y),
    ((1/a)*arcsinh(a*x))*t + (1-t)*y];
> Surface := plot3d(Helicat(x,y,1), x=-1..1,
    y=0..2*Pi/a, color=green):
> display(Surface);
```

Are the intermediate surfaces minimal?

Bibliography

A. Ancient Texts

Euclid, *Elements*, T.L. Heath, ed., New York: Dover, 1956.

Euclid, *Optics*, H. E. Burton, trans., *Journal of the Optical Society of America*, vol. 35, no. 5, pp. 357-372, 1945.

Euclid, *Phaenomena*, in *Euclidis opera omnia*, Heinrich Menge, ed., Lipsiae: B.G. Teubneri, 1883-1916.

The Holy Bible, NIV Zondervan Bible Publishers, 1985.

Khayyam, Omar, a paper (no title).
Translated in A. R. Amir-Moez, "A Paper of Omar Khayyam," *Scripta Mathematica*, 26(1963), pp.323-337.

Koran (Holy Qur-An), Abdullah Yusuf Ali, trans., New York: Harper Publishing, 1946.

Plato, *The Collected Dialogues*, Edith Hamilton and Huntington Carns, eds., Princeton, NJ: Bollinger, 1961.

AD. Art and Design

Edmondson, Amy C., *A Fuller Explanation: The Synergetic Geometry of R. Buckminster Fuller*, Boston: Birkhauser, 1987.

Ernst, Bruno, *The Magic Mirror of M. C. Escher*, New York: Random House, 1976.
A revealing look at the artist and the ideas behind his work.

Ghyka, Matila, *The Geometry of Art and Life*, New York: Dover Publications, 1977.

Henderson, Linda, *The Fourth Dimension and Non-Euclidean Geometry in Modern Art*, Princeton, NJ: Princeton University Press, 1983.

Linn, Charles, *The Golden Mean: Mathematics and the Fine Arts*, Garden City, NY: Doubleday, 1974.

Miyazaki, Kojiv, *An Adventure in Multidimensional Space*, New York: John Wiley and Sons, Inc., 1983.
"The art and geometry of polygons, polyhedra, and polytopes."

Williams, Robert, *The Geometrical Foundation of Natural Structure: A Source Book of Design*, New York: Dover, 1979.

An. Analysis

Bishop, Errett, and Douglas Bridges, *Constructive Analysis*, New York: Springer-Verlag, 1985.
The main book on constructive analysis.

Rudin, Walter, *Principles of Mathematical Analysis*, New York: McGraw Hill, 1964/1976.
For many years a standard text in analysis.

Strichartz, Robert S., *The Way of Analysis*, Boston: Jones and Bartlett Publishers, 1995.
"The presentation of the material in this book is often informal. A lot of space is given to motivation and a discussion of proof strategies."

DG. Differential Geometry

Berger, M., and B. Gostiaux, *Differential Geometry: Manifolds, Curves, and Surfaces*, New York: Springer-Verlag, 1988.

Casey, James, *Exploring Curvature*, Wiesbaden: Vieweg, 1996.
A truly delightful book full of "experiments" to physically explore curvature of curves and surfaces.

do Carmo, Manfredo, *Differential Geometry of Curves and Surfaces*, Englewood Cliffs, NJ: Prentice Hall, 1976.

Dodson, C. T. J., and T. Poston, *Tensor Geometry*, London: Pitman, 1979.
A very readable but technical text using linear (affine) algebra to study the local intrinsic geometry of spaces leading up to and including the geometry of the theory of relativity.

Dubrovin, B.A., A.T. Fomenko, S.P. Novikov, *Modern Geometry—Methods and Applications (Part I. The Geometry of Surfaces, Transformation Groups, and Fields)*, Robert G. Burns, trans., New York: Springer-Verlag, 1984.
A well-written graduate text.

Gauss, C.F., *General Investigations of Curved Surfaces*, Hewlett, NY: Raven Press, 1965.
A translation into English of Gauss' early papers on surfaces.

Gray, A, *Modern Differential Geometry of Curves and Surfaces*, CRC, 1993.
This is a very extensive book based on computations using Mathematica©.

Koenderink, Jan J., *Solid Shape*, Cambridge: M.I.T. Press, 1990.
Written for engineers and applied mathematicians, this is a discussion of the extrinsic properties of three-dimensional shapes. There are connections with applications and a nice section "Your way into the literature."

McCleary, John, *Geometry from a Differential Viewpoint*, Cambridge, UK: Cambridge University Press, 1994.
"The text serves as both an introduction to the classical differential geometry of curves and surfaces and as a history of ... the hyperbolic plane."

Millman, R.S., and G.D. Parker, *Elements of Differential Geometry*, Englewood Cliffs, NJ: Prentice-Hall, 1977.
A well-written text, which uses linear algebra extensively to treat the formalisms of extrinsic differential geometry.

Morgan, Frank, *Riemannian Geometry: A Beginner's Guide*, Boston: Jones and Bartlett, 1993.

Oprea, John, *Introduction to Differential Geometry and Its Applications*, Upper Saddle River: Prentice Hall, 1997.

Penrose, Roger, "The Geometry of the Universe," *Mathematics Today*, Lynn Steen, ed., New York: Springer-Verlag, 1978.
An expository discussion of the geometry of the universe.

Prakash, *Differential Geometry: An Integrated Approach*, New Delhi: Tata McGraw-Hill Publishing Company Limited, 1981.

Spivak, Michael, *A Comprehensive Introduction to Differential Geometry*, Wilmington, DE: Publish or Perish, 1979.
In five(!) volumes Spivak relates the subject back to the original sources. Volume V contains an extensive bibliography (up to 1979).

Stahl, Saul, *The Poincaré Half-Plane*, Boston: Jones and Bartlett Publishers, 1993.
This text is an analytic introduction to some of the ideas of intrinsic differential geometry starting from the Calculus.

Weeks, Jeffrey, *The Shape of Space*, New York: Marcel Dekker, 1985.
An elementary but deep discussion of the geometry on different two- and three-dimensional spaces.

DS. Dimensions and Scale

Abbott, Edwin A., *Flatland*, New York: Dover Publications, Inc., 1952.
A fantasy about two-dimensional beings in a plane encountering the third dimension.

Banchoff, Thomas and John Wermer, *Beyond the Third Dimension: Geometry, Computer Graphics, and Higher Dimensions*, New York: Springer-Verlag, 1983.

Burger, Dionys, *Sphereland*, New York: Thomas Y. Crowell Co., 1965.
A sequel to Abbott's *Flatland.*

Morrison, Phillip, and Phylis Morrison, *Powers of Ten: About the Relative Size of Things in the Universe*, New York: Scientific American Books, Inc., 1982.

Rucker, Rudy, *The Fourth Dimension*, Boston: Houghton Mifflin Co., 1984.
A history and description of various ways that people have considered the fourth dimension.

Rucker, Rudy, *Geometry, Relativity and the Fourth Dimension*, New York: Dover, 1977.

GC. Geometry in Different Cultures

Albarn, K., Jenny Miall Smith, Stanford Steele, Dinah Walker, *The Language of Pattern*, New York: Harper & Row, 1974.
An enquiry inspired by Islamic decoration.

Ascher, Marcia, *Ethnomathematics: A Multicultural View of Mathematical Ideas*, Pacific Grove, CA: Brooks/Cole, 1991.

Bain, George, *Celtic Arts: The Methods of Construction*, London: Constable, 1977.

Gerdes, Paulus, *Geometrical Recreations of Africa*, Maputo, Mozambique: African Mathematical Union and Higher Pedagogical Institute's Faculty of Science, 1991.

Kline, Morris, *Mathematics in Western Culture*, New York: Oxford University Press, 1961.

Pinxten, R., Ingrid van Dooren, Frank Harvey, *The Anthropology of Space*, Philadelphia: University of Pennsylvania Press, 1983.
Concepts of geometry and space in the Navajo culture.

Zaslavsky, Claudia, *Africa Counts*, Boston: Prindle, Weber, and Schmidt, Inc., 1973.
A presentation of the mathematics in African cultures.

Hi. History

Beckmann, Peter, *A History of* π , Boulder, CO: The Golem Press, 1970.
A well-written enjoyable book about all aspects of π.

Berggren, *Episodes in the Mathematics of Medieval Islam*, New York: Springer-Verlag, 1986.

Bold, Benjamin, *Famous Problems of Geometry and How to Solve Them*, New York: Dover Publications, Inc., 1969.

Calinger, Ronald, *Classics of Mathematics*, Englewood Cliffs, NJ: Prentice Hall, 1995.
Mostly a collection of original sources in Western mathematics.

Carroll, Lewis, *Euclid and His Modern Rivals*, New York: Dover Publications, Inc., 1973.
Yes! Lewis Carroll of *Alice in Wonderland* fame was a geometer. This book is written as a drama; Carroll has Euclid defending himself against modern critics.

Eves, Howard, *Great Moments in Mathematics (after 1650)*, Dolciani Mathematical Expositions, Vol. 7, Washington, DC: M.A.A., 1981.

Joseph, George, *The Crest of the Peacock*, New York: I.B. Tauris, 1991.
A non-Eurocentric view of the history of mathematics.

Kline, Morris, *Mathematical Thought from Ancient to Modern Times*, Oxford: Oxford University Press, 1972.
A complete Eurocentric history of mathematical ideas including differential geometry (mostly the analytic side).

Newell, Virginia K. (ed.) *Black Mathematicians and Their Works*, Ardmore, PA: Dorrance, 1980.

Richards, Joan, *Mathematical Visions*, Boston: Academic Press, 1988.
"The pursuit of geometry in Victorian England."

Seidenberg, A., "The Ritual Origin of Geometry," *Archive for the History of the Exact Sciences*, 1(1961), pp. 488-527.

Smeltzer, Donald, *Man and Number*, New York: Emerson Books, 1958.
History and cultural aspects of mathematics.

Valens, Evans G., *The Number of Things: Pythagoras, Geometry and Humming Strings*, New York: E.P. Dutton and Company, 1964.
This is a book about ideas and is not a textbook. Valens leads the reader through dissections, golden mean, relations between geometry and music, conic sections, etc.

LA. Linear Algebra and Geometry

Banchoff, T., and J. Wermer, *Linear Algebra through Geometry*, New York: Springer-Verlag, 1983.
For several years this was the text for Cornell's undergraduate linear algebra course.

Dodson, C. T. J., and T. Poston, *Tensor Geometry,* London: Pitman, 1979.
A very readable but technical text using linear (affine) algebra to study the local intrinsic geometry of spaces leading up to and including the geometry of the theory of relativity.

Fekete, Anton E., *Real Linear Algebra*, New York: Marcel Dekker, 1985.

Hannah, John, "A Geometric Approach to Determinants," *The American Mathematical Monthly*, 103(1996), pp.401-409.

Murtha, James A., and Earl R. Willard, *Linear Algebra and Geometry*, New York: Holt, Reinhart and Winston, Inc., 1966.
Includes affine and projective geometry.

Postnikov, M., *Lectures in Geometry: Semester II - Linear Algebra and Differential Geometry*, Moscow: Mir, 1982.

Taylor, Walter F., *The Geometry of Computer Graphics*, Pacific Grove, CA: Wadsworth and Brooks, 1992.

Mi. Minimal Surfaces

Hoffman, David, “The computer aided discovery of new embedded minimal surfaces,” *Mathematcal Intelligencer*, 9 (1987), pp.8-21.

Morgan, Frank, *Geometric Measure Theory: A Beginner’s Guide*, Boston: Academic Press, 1988.

Morgan, Frank, “Compound soap bubbles, shortest networks, and minimal surfaces,” AMS Video, Providence, RI: AMS, 1992.

Osserman, Robert, *A Survey of Minimal Surfaces*, 2nd edition, New York: Dover, 1986.

Osserman, Robert, “Minimal Surfaces in $\mathbb{R}^3$,” *Global Differential Geometry*, S. S. Chern, ed., MAA Studies in Mathematics, Volume 27, Washington, DC: MAA, 1989.

MP. Models, Polyhedra

Barnette, David, *Map Colouring, Polyhedra, and the Four-Colour Problem*, Dolciani Mathematical Expositions Vol. 8, Washington, DC: M.A.A., 1983.

Barr, Stephen, *Experiments in Topology*, New York: Crowell, 1964.

Cundy, M.H., and A.P. Rollett, *Mathematical Models*, Oxford: Clarendon, 1961.
Directions on how to make and understand various geometric models.

Lyusternik, L.A., *Convex Figures and Polyhedra*, Boston: Heath, 1966.

Row, T. Sundra, *Geometric Exercises in Paper Folding*, New York: Dover, 1966.
How to produce various geometric constructions merely by folding a sheet of paper.

Senechal, Marjorie, and George Fleck, *Shaping Space: A Polyhedral Approach*, Design Science Collection, Boston: Birkhauser, 1988.

Na. Nature

Cook, T.A., *The Curves of Life: Being an Account of Spiral Formations and their Applications to Growth in Nature, to Science, and to Art*, New York: Dover Publications, 1979.

Ghyka, Matila, *The Geometry of Art and Life*, New York: Dover Publications, 1977.

Mandelbrot, Benoit B., *The Fractal Geometry of Nature*, New York: W.H. Freeman and Company, 1983.
The book that started the popularity of fractal geometry.

McMahon, Thomas and James Bonner, *On Size and Life*, New York: Scientific American Library, 1983.
A geometric discussion of the shapes and sizes of living things.

Thom, Rene, *Structural Stabililty and Morphogenesis*, Redwood City, CA: Addison-Wesley, 1989.
A geometric and analytic treatment of "Catastrophe Theory."

Thompson, D'Arcy, *On Growth and Form*, Cambridge: Cambridge University Press, 1961.
A classic on the geometry of the natural world.

NE. Non-Euclidean Geometries (Mostly Hyperbolic)

Greenberg, Marvin J., *Euclidean and Non-Euclidean Geometries: Development and History*, New York: Freeman, 1980
This is a very readable textbook that includes some philosophical discussions.

Millman, Richard S., and George D. Parker, *Geometry: A Metric Approach with Models*, New York: Springer-Verlag, 1981.
A modern formal axiomatic approach.

Nikulin, V.V., and I.R. Shafarevich, *Geometries and Groups*, Berlin: Springer-Verlag, 1987.
Using transformation groups to study spherical, hyperbolic, and toroidal geometries.

Petit, Jean-Pierre, *Euclid Rules OK? The Adventures of Archibald Higgins*, London: John Murray, 1982.
A pictorial, visual tour of non-Euclidean geometries.

Ryan, Patrick J., *Euclidean and Non-Euclidean Geometry: An Analytic Approach*, Cambridge: Cambridge University Press, 1986.

Schwerdtfeger, Hans, *Geometry of Complex Numbers: Circle Geometry, Moebius Transformation, Non-Euclidean Geometry*, New York: Dover Publications, Inc., 1979.

Ph. Philosophy

Benacerraf, Paul, and Hilary Putman, *Philosophy of Mathematics: Selected Readings*, Cambridge: Cambridge University Press, 1964.

Hofstadter, Douglas R., *Gödel, Escher, Bach: An Eternal Golden Braid*, New York: Basic Books, 1979.

Lachterman, David Rapport, *The Ethics of Geometry: A Genealogy of Modernity*, New York: Routledge, 1989.

Lakatos, I., *Proofs and Refutations*, Cambridge: Cambridge University Press, 1976.

Stein, Charles (ed.), *Being = Space X Action*, IO Vol. 41, Berkeley, CA: North Atlantic Books, 1988.
"Searches for Freedom of Mind through Mathematics, Art, and Mysticism."

Tymoczko, Thomas, *New Directions in the Philosophy of Mathematics*, Boston: Birkhauser, 1986.

RN. Real Numbers

Conway, John Horton, *On Numbers and Games*, New York: Academic Press, 1976.
Conway presents a version of the real numbers that has come to be called "Conway Numbers," which are too large to be a set even though, as the title suggests, they have found applications in game theory.

Ebbinghaus et al, *Numbers*, New York: Springer-Verlag, 1991.
A lively story about the concept of number.

Epstein, Richard L., and Walter A. Carnielli, *Computability: Computable Functions, Logic, and the Foundations of Mathematics*, Pacific Grove, CA: Wadsworth & Brooks/Cole, 1989.
"This book...deals with a very basic problem: What is computable?"

Laugwitz, "Infinitely Small Quantities in Cauchy's Textbooks," *Historia Mathematica*, 14 (1987), pp.258-274.

Moore, Ramon, *Methods and Applications of Interval Analysis*, Philadelphia: Society of Industrial and Applied Mathematics, 1979.
Interval analysis "an approach to computing that treats an interval as a new kind of number."

Simpson, "The Infidel Is Innocent," *The Mathematical Intelligencer*, 12 (1990), pp.42-51.
An accessible expositon of the non-standard reals.

Turner, Peter R., "Will the 'Real' Real Arithmetic Please Stand Up?", *Notices of the AMS*, **38** (1991), pp.298-304.
An article about various finite representations of real numbers used in computing.

SP. Spherical and Projective Geometry

Henderson, David W., *Experiencing Geometry on Plane and Sphere*, Upper Saddle River, NJ: Prentice Hall, 1996.

Todhunter, Isaac, *Spherical Trigonometry*, London: Macmillan, 1886.
All you want to know, and more, about trigonometry on the sphere. Well written with nice discussions of surveying.

Whicher, Olive, *Projective Geometry: Creative Polarities in Space and Time*, London: Rudolf Steiner Press, 1971.
Projective geometry is the geometry of perception and prospective drawings.

SG. Symmetry and Groups

Budden, F.J., *Fascination of Groups*, Cambridge: Cambridge University Press, 1972.
This is a fascinating book that relates algebra (groups) to geometry, music, and so forth, and has a nice description of symmetry and patterns.

Bunch, Bryan H., *Reality's Mirror: Exploring the Mathematics of Symmetry*, New York: John Wiley, 1989.

Burn, R.P., *Groups: A Pathway to Geometry*, Cambridge: Cambridge University Press, 1985.

Weyl, Hermann, *Symmetry*, Princeton, NJ: Princeton University Press, 1952.
A readable discussion of all mathematical aspects of symmetry especially its relation to art and nature—nice pictures. Weyl is a leading mathematician of this century.

SE. Surveys and General Expositions

Davis, P.J., and R. Hersh, *The Mathematical Experience*, Boston: Birkhauser, 1981.
A very readable collection of essays by two present-day mathematicians. I think every mathematics major should own this book.

Ekeland, Ivar, *Mathematics and the Unexpected*, Chicago: University of Chicago Press, 1988.

Gaffney, Matthew P. and Lynn Arthur Steen, *Annotated Bibliography of Expository Writing in the Mathematical Sciences*, Washington, DC: M.A.A., 1976.

Gamow, George, *One, Two, Three ... Infinity*, New York: Bantam Books, 1961.
A well-written journey through mathematical ideas.

Guillen, Michail, *Bridges to Infinity: The Human Side of Mathematics*, Los Angeles: Jeremy P. Tarcher, 1983.

Hilbert, David, and S. Cohn-Vossen, *Geometry and the Imagination*, New York: Chelsea Publishing Co., 1983.
They state "it is our purpose to give a presentation of geometry, as it stands today [1932], in its visual, intuitive aspects." It includes an introduction to differential geometry, symmetry, and patterns (they call it "crystallographic groups"), and the geometry of spheres and other surfaces. Hilbert is the most famous mathematician of the first part of this century.

Honsberger, Ross, *Mathematical Gems*, Dolciani Mathematical Expositions, Vol. 2, Washington, DC: M.A.A., 1973.

Honsberger, Ross, *Mathematical Gems II*, Dolciani Mathematical Expositions, Vol. 4, Washington, DC: M.A.A., 1976.

Honsberger, Ross, *Mathematical Morsels*, Dolciani Mathematical Expositions, Vol. 1, Washington, DC: M.A.A., 1978.

Honsberger, Ross, *Mathematical Plums*, Dolciani Mathematical Expositions, Vol. 4, Washington, DC: M.A.A., 1979.
Expository stories about mathematics.

Jester, Norton, *The Dot and the Line: A Romance in Lower Mathematics*, New York: Random House, 1963.
A mathematical fable.

Lang, Serge, *The Beauty of Doing Mathematics: Three Public Dialogues*, New York: Springer-Verlag, 1985.
Expository work by a famous mathematician.

Lieber, Lillian R., *The Education of T.C. Mits (The Celebrated Man in the Street)*, New York: W.W. Norton, 1972.
A mathematical fantasy.

Péter, Rozsa, *Playing with Infinity*, New York: Dover Pubslishing, Inc., 1961.
"Mathematical explorations and excursions."

Steen, Lynn Arthur (ed.), *Mathematics Tomorrow*, New York: Springer-Verlag, 1981.
Expository essays.

Steen, Lynn Arthur (ed.), *Mathematics Today: Twelve Informal Essays*, New York: Springer-Verlag, 1978.

Stewart, Ian, *The Problems of Mathematics*, Oxford: Oxford University Press, 1987.

Tp. Topology

Francis, G.K., *A Topological Picturebook*, New York: Springer Verlag, 1987.
Francis presents elaborate and illustrative drawings of surfaces and provides guidelines for those who wish to produce such drawings.

Hurewics, W., and Wallman, H., *Dimension Theory*, Princeton: Princeton University Press, 1941.
Contains a proof of the Invariance of Domain in the context of the theory of the dimension of topological spaces.

Newman, M.H.A., *Elements of the Topology of Plane Sets of Points*, Cambridge: Cambridge University Press, 1964.
Contains a proof of the Invariance of Domain that is the most geometric.

Spanier, *Algebraic Topology*, New York: McGraw Hill Book Company, 1966.
Contains a proof of the Invariance of Domain based on algebraic topology.

Tx. Geometry Texts

Coxeter, H.S.M., *Introduction to Geometry*, New York: Wiley, 1969.
This is a collection of diverse topics including non-Euclidean geometry, symmetry, patterns, and much, much more. Coxeter is one of the foremost living geometers.

Eves, Howard, *A Survey of Geometry,* Vol. 1, Boston: Allyn & Bacon, 1963.

Eves, Howard, *Modern Elementary Geometry*, Boston: Jones and Bartlett Publishing, 1992.

Henderson, David W., *Experiencing Geometry on Plane and Sphere*, Upper Saddle River, NJ: Prentice Hall, 1996.

Jacobs, Harold R., *Geometry*, San Fransisco: W.H. Freeman and Co., 1974.
A high-school-level text based on guided discovery.

Serra, Michael, *Discovering Geometry: An Inductive Approach*, Berkeley, CA: Key Curriculum Press, 1989.

Z. Miscellaneous

Davis, Phillip, *The Thread: A Mathematical Yarn*, Boston: Birkhäuser, 1983.

Ho, Chung-Wu, "Decomposition of a Polygon into Triangles," *Mathematical Gazette*, 60 (1976), pp.132-134.

Kempe, A.B., *How to Draw a Straight Line*, London: Macmillan, 1877.

Pottmann, Helmut, "Rational curves and surfaces with rational offsets," *Computer Aided Geometric Design*, 12 (1995), pp.175-192.

Sah, C.H., *Hilbert's Third Problem: Scissors Congruence*, London: Pitman, 1979.

Notation Index

Subject Index

H

I

L

M

N

O

P

Q

R

S

T

V

W

Z